MIDI Systems and Control

MIDI Systems and Control

Second Edition

Francis Rumsey

Focal Press
An imprint of Butterworth-Heinemann Ltd
Linacre House, Jordan Hill, Oxford OX2 8DP

A member of the Reed Elsevier plc group

OXFORD   LONDON   BOSTON
MUNICH   NEW DELHI   SINGAPORE   SYDNEY
TOKYO   TORONTO   WELLINGTON

First published 1990
Second edition 1994
Reprinted 1995

British Library Cataloguing in Publication Data
A catalogue record for this book is
available from the British Library

Library of Congress Cataloguing in Publication Data
A catalogue record for this book is
available from the Library of Congress

ISBN 0 240 51370 3

Cover picture kindly supplied by Yamaha-Kemble Music (UK) Ltd

Typesetting and layout by Designer Publishing, Guildford
Printed and bound in Great Britain by Clays Ltd, St Ives plc

Contents

Preface

MIDI Systems and Control is a book about how MIDI, the Musical Instrument Digital Interface, works. This is its second edition, and it has been written firstly because the publisher has sold out of the first edition (always a good sign!) and secondly because the field of MIDI control has grown out of almost all recognition since I first started putting the book together in 1988. The book has been almost entirely rewritten for this edition, and an opening chapter has been added as an introduction to computer systems and terminology for those who need it. There are many more diagrams, and in all the book is roughly twice the size of the original edition.

I originally wrote this book in an attempt to *explain* the MIDI standard and show how it had been implemented in practical systems. I have maintained that principle in the second edition, having brought the coverage right up to date, and I have added information on ways in which MIDI may be integrated with digital audio and video systems. This is a book for anyone who wants to understand the principles of MIDI control, and that includes people who use MIDI equipment as well as those who want to design devices. It is not, however, a book about programming, because there are many alternative publications which have covered this subject very adequately, and it is not a book about how to use MIDI equipment. Neither is it a book about commercial hardware and software, since the product line changes too rapidly to make that a practical proposition.

The first edition was often called 'a readable version of the MIDI standard', but it is not intended to be used as a substitute for referring to the standards documents themselves, the source of which is detailed at the end of the book. Indeed there are a number of places in this book in which it has not been considered appropriate to describe all of the possible messages or protocols, because the objective has been to explain concepts. The job of standards documents is to lay out the *status quo* in formal language, and those wishing to implement MIDI control should use this book in conjunction with the standards documents, treating the book as an explanation of and commentary upon the information contained therein.

It is hard to imagine what the music industry would be like today without MIDI control. The formulation and support of a universal, worldwide standard some ten years ago has effectively created an industry of its own, and, although there is sometimes talk of replacing MIDI with something faster and better, it is unquestionable that it will continue to be the principal means by which computer music equipment and its related peripherals are controlled for the foreseeable future.

Francis Rumsey,
December 1993

Chapter 1

An introduction to computer systems and terminology

The whole field of MIDI control is directly related to computer technology. MIDI is a serial remote control interface for musical instruments and other equipment, and therefore if one is to understand MIDI it is vital also to have at least an appreciation of computer systems, the role of software, the terminology used in remote control interfaces, and the factors influencing the performance of digital computer systems. Such an understanding will help those who use MIDI hardware and software to appreciate what is happening inside the equipment, and may also assist when planning systems, purchasing computers and working with operating systems. For this reason the first chapter of this book is designed to introduce these principles to those without a background in computing or information technology, in order that the rest of the book will be more understandable. Those who feel comfortable with such material can ignore this chapter and begin with Chapter 2, where coverage of MIDI itself begins.

1.1 MIDI and its relationship to computer technology

MIDI is an acronym for the Musical Instrument Digital Interface. The interface transmits and receives remote control commands for musical instruments and other studio devices in a digital form – that is in the form of binary data, transmitted electrically. The commands transmitted from one MIDI device to another indicate actions to be taken by the controlled device, such as turning on a note or altering an expression parameter. The result, in the case of a musical instrument, will normally be the generation of sound or the altering of a sound. In this way, MIDI commands control the generation of sound by remote instruments, but the MIDI interconnection does not normally carry sound information itself. MIDI is therefore different from digital audio (whereby sounds themselves are converted into a digital form and stored or transmitted). MIDI replaces the analogue communication methods used over ten years ago to control musical instruments, described in more detail in section 2.1.

MIDI equipment relies on microcomputers. In fact virtually all MIDI devices contain at least one microprocessor and its associated peripherals to manage the process of interpreting the commands which arrive and their execution. As many will be aware, a microcomputer is not just a device with a screen and keyboard which sits on the desktop, used for typing letters and working out the budget, a microcomputer is fundamentally a system designed to manipulate binary numerical information (data), perhaps store the data, and communicate with peripheral devices in order to accept commands or give out information. The personal desktop computer takes in commands from a keyboard and perhaps a mouse, is able to store it in various places such as in solid state memory or on a disk, and (by virtue of the software program running on the microprocessor) may perform operations on stored information in order to display a result or control a process.

A washing machine may contain a simple microcomputer designed to control a mechanical process. Its inputs will be from the front panel, and also from the various sensors built into the machine to tell it when the motors are rotating, which valves are open, whether the water pressure is high enough and so on. Its outputs will be to the display, and to the motors, relays and valves which initiate the various washing processes. The computer within a MIDI controlled sound generator may be very similar, but will be programmed to perform different functions. Its job will be to accept commands from any musical keyboard present, from the front panel controls, and also from the MIDI input. Its outputs will be to the sound generators themselves, perhaps to a small display, and to the MIDI output. Essentially therefore the microcomputer has inputs, outputs, processing and storage. The concepts are illustrated in Figure 1.1, and are expanded upon further in section 1.6.

The computer communicates with the outside world by means of interfaces or ports. These interfaces transmit and/or receive data, normally in electrical form (see section 1.6.4). The speed at which the data is passed through an interface depends on the application. In washing machine control there is normally no need for high speed communication of information between the various controls and the mechanism, since events take place relatively infrequently and are not time critical (at least not in the computing sense). Provided that the shirts stop spinning at roughly the right time the machine will work satisfactorily. With the desktop computer information may be arriving and being sent out at a wide range of different rates. From the keyboard information will arrive relatively slowly, since even the fastest typist will only be generating a few characters per second, and thus the interface connecting the keyboard could be relatively slow. The interface connecting the disk drive to the computer, though, could usefully be much faster, since the faster one can store data on a disk the better.

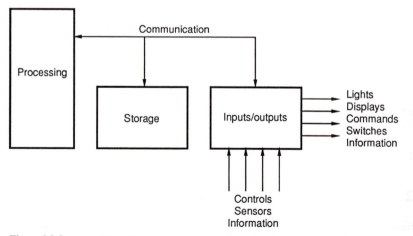

Figure 1.1 Conceptually, a microcomputer consists of inputs and outputs, processing and storage

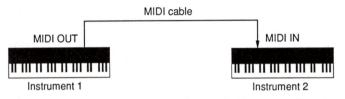

Figure 1.2 Simple MIDI interconnection of two musical instruments allows Instrument 2 to imitate Instrument 1

Music lends itself neatly to computer storage and processing, because musical compositions generally consist of a number of defined events using defined sounds which can be represented by specific binary codes. The speed with which musical events occur depends on the composer or player, and any interface designed for use with musical information needs to be fast enough to transmit the events as they occur without a noticeable delay. A device receiving that information must be able to act on the information and generate musical sounds so that they occur at the intended time. Taking the simple situation shown in Figure 1.2, where two musical instruments are connected together using an as yet unspecified MIDI interface, one of the aims is that the receiving instrument will imitate the transmitting device by acting on the received commands as the music is played, with no noticeable delay. What is really needed is an interface capable of handling 'real-time' musical commands. Now in reality commands will take a finite time to be sent from one device to the other, so the practical solution is to design an interface whose speed of transmission is sufficiently high that any delay which might result will be insignificant from a musical point of view. As will be seen later on, the MIDI interface is a so-called 'serial interface', whereby information for more than one device may be transmitted over a single communications channel, requiring careful optimisation of the data flow in order to maintain the impression of real-time operation.

Music computer software, as well as forming the hidden operating system of a MIDI-controlled device, is available for a wide range of desktop computers. Once musical information and other remote control data is coded in a digital form and stored in a memory, it can easily be changed. This allows musical compositions to be stored in a computer memory and manipulated in a wide variety of different ways. Effects which would either have been impossible with conventional musical techniques, or which would otherwise have taken a vast amount of time and equipment, become possible. There are parallels here with digital audio and video processing in which the data representing sound and pictures may be manipulated by effects software to change the result out of all recognition. Most people are familiar with the visual effects gymnastics performed on television, which are simply the result of taking the binary data representing the picture elements and rearranging them, modifying them, processing them and using them to create altered duplicates. Musical events can be treated in a similar way, leaving it only up to the imagination of the composer and the flexibility of the software to determine what is possible. More will be said about this later.

1.2 Analogue and digital information

Before going any further it is necessary to compare analogue and digital information. The human senses deal mainly with analogue information, but computers deal

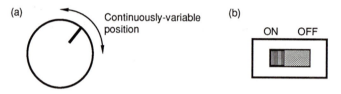

Figure 1.3 (a) A continuously variable control such as a rotary knob is an analogue controller. (b) A two-way switch is a digital controller

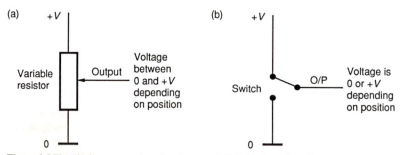

Figure 1.4 Electrical representation of analogue and digital information. The rotary controller of Figure 1.3(a) could adjust a variable resistor, producing a voltage anywhere between the limits of 0 and $+V$, as shown in (a). The switch connected as shown in (b) allows the selection of either 0 or $+V$ states at the output

internally with digital information, and thus there is the need for conversion between one domain and the other at various points.

Analogue information is made up of a continuum of values, which at any instant may have any value between the limits of the system. For example, a rotating knob may have one of an infinite number of positions – it is therefore an analogue controller (see Figure 1.3). A simple switch, on the other hand, can be considered as a digital controller, since it has only two positions – off or on. It cannot take any value in between. The brightness of light which we perceive with our eyes is analogue information, and as the sun goes down the brightness falls gradually and smoothly, whereas a household light without a dimmer may be either on or off – its state is binary (that is it has only two possible states). A single item of binary information is called a bit (binary digit), and a bit can only have the value one or zero (corresponding, say, to high and low, or on and off states of the electrical signal).

Electrically, analogue information may be represented as a varying voltage or current. If the rotary knob of Figure 1.3 is connected to a variable resistor and a voltage supply, its position will affect the output voltage (see Figure 1.4) which, like the knob's position, may have any value between the limits – in this case anywhere between zero volts and $+V$. The switch may be connected to a similar voltage supply, and in this case the output voltage can only be either zero volts or $+V$. In other words the electrical information which results is binary. The high $(+V)$ state could be said to correspond to a binary one, and the low state to binary zero (although in many real cases it is actually the other way around). One switch can represent only one binary digit (or bit), but most digital information is made up of more than one bit, allowing digital representations of a number of fixed values.

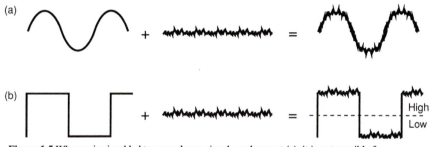

Figure 1.5 When noise is added to an analogue signal, as shown at (a), it is not possible for a receiver to know what is the original signal and what is the unwanted noise. With the binary signal, as shown at (b), it is possible to extract the original information even when noise has been added. Everything above the decision level is high, and everything below it is low

Analogue information in an electrical form is converted into a digital electrical form using a device known as an analogue-to-digital (A/D) convertor – indeed it must be if it is to be handled by any logical system such as a computer. This process will be described in section 1.5. The output of an A/D convertor is a binary numerical value representing as accurately as possible the analogue voltage which was converted.

Digital information made up of binary digits is inherently more resilient to noise and interference than analogue information, as shown in Figure 1.5. If noise is added to an *analogue* signal then it becomes very difficult to tell at any later stage in the signal chain what is the wanted signal and what is the unwanted noise, since there is no means of distinguishing between the two. If noise is added to a digital signal it *is* possible to extract the important information at a later stage, since it is known that only two states matter – the high and low, or one and zero states. By comparing the signal amplitude with a fixed decision point it is possible for a receiver to decide that everything above the decision point is 'high', whilst everything below it is 'low'. Any levels in between can be classified in the nearest direction. Thus for any noise to influence the state of a digital signal, it must be at least large enough in amplitude to cause a high level to be interpreted as 'low', or vice versa.

The timing of digital signals may also be corrected to some extent, using a similar method, which gives digital signals another advantage over analogue. If the timing of bits in a digital message becomes unstable, such as after having been passed over a long cable, resulting in timing 'jitter', the signal may be reclocked at a stable rate ensuring that the timing stability of the information is restored.

1.3 Binary number systems

In the decimal number system, each digit of a number represents a power of ten. In a binary system each digit or bit represents a power of two (see Figure 1.6). It is possible to calculate the decimal equivalent of a binary integer (whole number) by using the method shown. A number made up of more than one bit is called a binary 'word', and an 8 bit word is called a 'byte' (from 'by eight'). Four bits is called a 'nibble'. The more bits there are in a word the larger the number of states it can represent, with eight bits allowing 256 (2^8) states and sixteen bits allowing 65 536 (2^{16}). The bit with the lowest weight (2^0) is called the least significant bit or LSB,

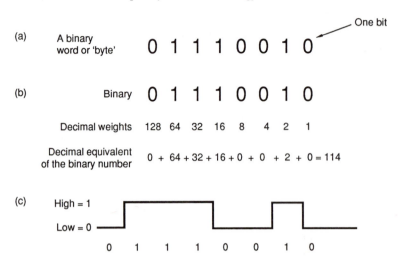

(a)   A binary word or 'byte'   0 1 1 1 0 0 1 0 ← One bit

(b)   Binary   0 1 1 1 0 0 1 0

Decimal weights   128 64 32 16 8 4 2 1

Decimal equivalent of the binary number   0 + 64 + 32 + 16 + 0 + 0 + 2 + 0 = 114

(c)   High = 1

Low = 0

0 1 1 1 0 0 1 0

Figure 1.6 (a) A binary number (word or 'byte') consists of a number of bits. (b) Each bit represents a power of two. (c) Binary numbers can be represented electrically in pulse-code modulation (PCM) by a string of high and low voltages

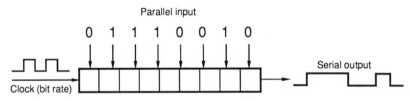

Parallel input

0 1 1 1 0 0 1 0

Serial output

Clock (bit rate)

Figure 1.7 A shift register is used to convert a parallel binary word into a serial format. The clock is used to shift the bits one at a time out of the register, and its frequency determines the data rate. The data may be clocked out of the shift register either MSB or LSB first, depending on the device and its configuration

and that with the greatest weight is called the most significant bit or MSB. The term kilobyte or kbyte is used to mean 1024 or 2^{10} bytes, and the term megabyte or Mbyte represents 1024 kbytes.

Electrically it is possible to represent a binary word in either 'serial' or 'parallel' form. In serial communication only one connection need be used, and the word is clocked out one bit at a time using a device known as a shift register. The shift register is previously loaded with the word in parallel form (see Figure 1.7). The rate at which the serial data is transferred depends on the speed of the clock. In parallel communication, each bit of the word is transferred over a separate connection. Further coverage is given of serial and parallel communication in section 1.8.5.

Because binary numbers can become fairly unwieldy when they get long, various forms of shorthand are used to make them more manageable. The most common of these is hexadecimal. The hexadecimal system represents decimal values from 0 to 15 using the sixteen symbols 0–9 and A–F, according to Table 1.1, thus each hexadecimal digit corresponds to four bits or one nibble of the binary word. An example showing how a long binary word may be written in hexadecimal (hex) is

Table 1.1 Hexadecimal and decimal equivalents to binary numbers

Binary	Hexadecimal	Decimal
0000	0	0
0001	1	1
0010	2	2
0011	3	3
0100	4	4
0101	5	5
0110	6	6
0111	7	7
1000	8	8
1001	9	9
1010	A	10
1011	B	11
1100	C	12
1101	D	13
1110	E	14
1111	F	15

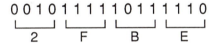

Figure 1.8 This 16 bit binary number may be represented in hexadecimal as shown, by breaking it up into 4 bit chunks (nibbles) and representing each chunk as a hex digit

shown in Figure 1.8 – it is simply a matter of breaking the word up into 4 bit chunks and converting each chunk to hex. Similarly, a hex word can be converted to binary by using the reverse process.

Negative integers are often represented in a form known as 'twos complement'. To obtain the twos complement or negative version of a positive number it is necessary to invert all the bits and add one. The advantage of twos complement representation is that binary arithmetic can be performed on positive and negative numbers, and the result will be correct in twos complement form. The MSB of negative twos complement numbers is always a '1'.

1.4 Logical operations

Most of the processing operations that occur within a computer can be boiled down to a fast sequence of simple logical operations. The apparent 'power' of the computer and its ability to perform complex tasks are really due to the speed with which simple operations are performed.

The basic family of logical operations is shown in Figure 1.9 in the form of a truth table next to the electrical symbol which represents each 'logic gate'. The AND operation gives an output only when both its inputs are true; the OR operation gives an output when either of its inputs are true; and the XOR (exclusive OR) gives an output only when one of its inputs is true. The invertor or NOT gate gives an output which is the opposite of its input, and this is often symbolised using a small circle on inputs or outputs of devices to indicate inversion.

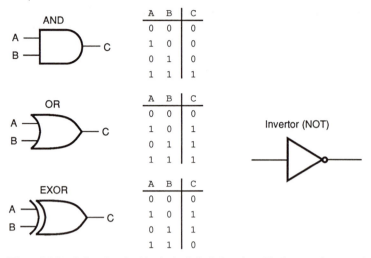

Figure 1.9 Symbols and truth tables for basic logic functions. The invertor shown on the right has an output which is always the opposite of the input. The circle on the invertor's output can be used to signify inversion on any input or output of a logic gate

1.5 Analogue-to-digital and digital-to-analogue conversion

It is not intended to cover this subject in detail, but the basic principles will be given. Conversion is a very large subject in its own right, but enough information will be given here to enable the reader to understand what happens when an analogue control's position is digitised. The process of sound digitising is discussed in more detail in Chapter 3.

As already stated, an analogue signal can have an infinite number of amplitudes, whereas a digital signal can only have a certain number of fixed values. The number of fixed values possible with a digital signal depends on the number of bits in the binary words involved, as previously described. In order to convert an analogue signal into a digital signal it is necessary to measure its amplitude at specific points in time (called 'sampling'), and to assign a binary value to each measurement (called 'quantising'). The diagram in Figure 1.10 shows the rotary knob used earlier against a fixed scale running from 0 to 9. If one were to quantise the position of the knob it would be necessary to determine which point of the scale it was nearest, and unless the pointer was at exactly one of the increments the quantising process would involve a degree of error. It will be seen that the maximum error is actually plus or minus half of an increment, since once the pointer is more than halfway between one increment and the next it will be quantised to the next.

Quantising error is an inevitable side effect in the process of A/D conversion, and the degree of error depends on the quantising scale used. Considering binary quantisation, a 4 bit scale offers sixteen possible steps, an 8 bit scale offers 256 steps, and a 16 bit scale 65 536. The more bits, the more accurate the process of quantisation. In MIDI systems, analogue controllers such as sliders and rotary controls, are often only quantised to seven bits (for reasons which will become apparent), but this is often adequate for many forms of control where relatively coarse changes are required. 14 bit quantisation is used for controllers which need finer resolution.

In older systems, the position of an analogue control is first used to derive an analogue voltage (as shown earlier in Figure 1.4), and then that voltage is converted into a digital value using an A/D convertor (see Figure 1.11). More recent controls may be in the form of binary encoders whose output is already digital. Unlike analogue controls, switches do not need the services of an A/D convertor for their outputs to be useable by a computer, since the switch's output is normally binary in the first place. Only one bit is needed to represent the position of a simple switch.

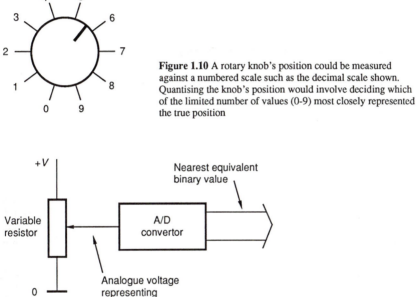

Figure 1.10 A rotary knob's position could be measured against a numbered scale such as the decimal scale shown. Quantising the knob's position would involve deciding which of the limited number of values (0-9) most closely represented the true position

Figure 1.11 In older equipment, a control's position was digitised by sampling and quantising an analogue voltage derived from a variable resistor connected to the control knob

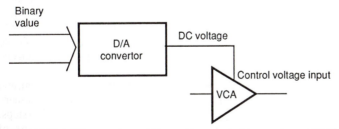

Figure 1.12 A D/A convertor could be used to convert a binary value representing a control's position into an analogue voltage. This could then be used to alter the gain of a voltage-controlled amplifier (VCA)

The rate at which switches and analogue controls are sampled depends very much on how important it is that they are updated regularly. Many analogue mixing consoles sample the positions of automated controls once per television frame (40 ms in Europe), whereas some digital mixers sample controls as often as once per audio sample period (roughly 20 μs). Clearly the more regularly a control is sampled the more data will be produced, since there will be one binary value per sample.

Digital-to-analogue conversion is the reverse process, and involves taking the binary value which represents one sample and converting it back into an electrical voltage. In a control system this voltage could then be used to alter the gain of a voltage-controlled amplifier (VCA), for example, as shown in Figure 1.12. Alternatively it may not be necessary to convert the word back to an analogue voltage at all. Many systems are entirely digital and can use the binary value derived from a control's position as a multiplier in a digital signal processing operation. A signal processing operation may be designed to emulate an analogue control process.

1.6 Basic computer system principles

This section contains an overview of the functions and operation of the key devices in a microcomputer system. Although practical implementations obviously differ considerably from each other, the principles remain much the same. Also, although computer systems become more complex all the time, it is still necessary to understand how a simple one works. The concepts involved in more complex computers are often just extensions of these basic ideas.

1.6.1 Buses

Figure 1.13 shows a basic block diagram of a microcomputer system, showing the main functional blocks. Before looking at specific devices within this system it is important to understand the function of a bus, which is normally a parallel collection of wires or printed circuit tracks each carrying one bit of a binary word. Thus an 8 bit data bus will carry data eight bits at a time in parallel, and a 16 bit address bus will carry sixteen address bits in parallel.

Information may be made to appear to travel in either direction on a bus, although direction is not really the right way of imagining it. The apparent direction of data flow simply depends on which device is placing the data on the bus and which device

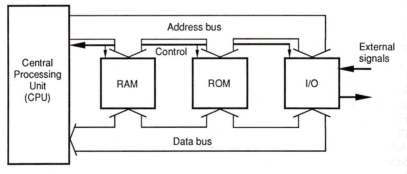

Figure 1.13 Simple block diagram of a microcomputer (see text)

is reading it. The bus is normally shared between a number of devices, and data is routed between the CPU and a device by enabling the appropriate receiver or transmitter at the appropriate time. Clearly it is important to ensure that two devices do not attempt to place data on the bus at the same time.

1.6.2 The CPU

The central processing unit (CPU) is effectively the system controller, and its main functions are to sequence and interpret instructions which are fetched from memory, to perform logical operations on whole binary words, to store data temporarily and to monitor external requests for attention (interrupts). The CPU communicates with other devices using a data bus, an address bus and various control lines, as outlined in section 1.6.3. The CPU is connected to a crystal clock which generates a synchronisation signal at a rate of a number of megahertz. This clock is the 'driving force' behind the whole sequence of operations that occurs in the computer, because it is the changes of the clock signal which instigate the next event in the logical program sequence. Built into the CPU is a sequencer (no direct relation to the MIDI sequencer) which determines the order of logical events in the many gates, stores and counters of the CPU. This sequencer is programmed by an instruction decoder which reads the binary words that form the instructions as they are fetched from memory. There is only a limited number of possible commands that are understood by the CPU, and these commands are called the instruction set. On each cycle of the clock the sequencer steps one stage further through the sequence programmed by the last fetched instruction.

One of the most important devices within the CPU is the ALU (arithmetic and logic unit), which is a programmable device designed to perform logical/mathematical operations on the data. An ALU is really a large collection of gates of the type described above, and typically has two inputs which are a number of bits wide, and a number of control inputs and outputs. The control inputs are used to determine what logical operation will be performed, and 'carry in' and 'carry out' lines are used when the result of an operation is too large or too small and so overflows the MSB of the word. One of the inputs to the ALU is normally fed from a temporary store known as the accumulator, and the other input may be data read in over the data bus, for example.

The CPU contains a program counter which normally starts from zero at power up, and whose output is routed to the address bus in order to point to the location of the next command stored in the memory. When the clock has run the appropriate number of cycles (known as machine cycles) to run the sequencer through the current instruction, the program counter is incremented to the next address and the next instruction is fetched from memory. A typical sequence of events for a single instruction might go something like this:

Place next instruction address onto address bus
Read contents of that memory location into instruction decoder
(Decode instruction to determine next step)
Fetch next byte of data from memory
Place that byte at one input to the ALU
Add that byte to the byte contained in the accumulator
Store the result in a temporary location
Increment program counter

This is one instruction being executed, but it has taken a number of cycles of the clock.

The CPU also contains a temporary store called the stack, which is configured in the last-in–first-out mode. The stack is like a sprung plate holder, such as might be found in a cafeteria, which holds a pile of plates. The last plate to be pushed on to the top of the pile is the first one to come off the top if someone wants a plate. Data is sometimes pushed temporarily onto the stack, using it as a holding place while another action is executed, whereafter the data is pulled back off the stack.

The instruction set of a typical CPU contains commands such as those to move data from one location to another, those to perform mathematical operations on a pair of numbers and those to jump the program execution to a new memory address. There are also important commands to read and write data to input/output (I/O) ports, which is the means by which data is communicated to and from the outside world – MIDI interfaces included. This is covered further in section 1.6.4.

1.6.3 ROM and RAM

There are two main types of solid-state memory in the typical computer: read only memory (ROM), and random access memory (RAM). (Solid-state devices are normally based on small slivers or 'chips' of silicon which have been engineered to behave as very large collections of electronic devices such as transistors, resistors and capacitors. Tiny wires are attached to discrete points on the silicon wafer, and these are connected to the pins on the chip packaging so as to allow connection to the outside world.)

There are many sub-types of ROM and RAM with subtle differences. All memory devices store particular bit patterns in different locations, known as addresses. These may be imagined as a set of 'pigeonholes', each with a unique matrix reference, as illustrated in Figure 1.14. These memory addresses may hold data that is anything from one bit wide to many. In the case of 1 bit memory chips it may be necessary to configure them in an array in order to store 8 bit words.

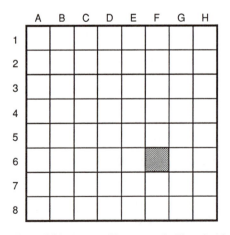

Figure 1.14 Memory addresses may be likened to the 'grid references' of locations in this matrix. The shaded square has the address F6

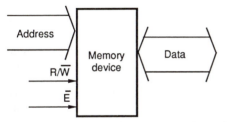

Figure 1.15 A memory device connected to data and address buses. The R/W line determines the direction of data flow (high for read, low for write). The enable line enables this particular device to read or write

Memory is characterised in block diagram form as shown in Figure 1.15, showing an address bus, a data bus, an enable line and a read/write (R/W) line (ROM has no R/W line because it is read-only). In order to write a byte of data to the memory, an address is presented on the address bus and the appropriate data is presented on the data bus. The R/W line is changed to the write state, and the memory chip concerned is enabled by holding the enable line in the appropriate state. Read operations are performed in a similar manner, except with the R/W line in the opposite state.

ROM is designed for data which will not be changed by the system, such as parts of the operating system software (the commands which program the CPU). True ROM is programmed by the manufacturer by blowing tiny fusible links in the chip at certain memory addresses in order to set particular bit patterns. These are then permanently stored and cannot be erased, thus making true ROM a good permanent store for limited amounts of data (up to a few megabytes). The data is retained even with no power to the chip. Other forms of ROM exist which may be modified with varying degrees of ease. EPROM is erasable programmable ROM, and may be erased by exposing a small window to ultraviolet light for around 20 minutes, after which it may be reprogrammed in a so-called 'PROM blower' by applying a higher than normal voltage to the pins of the EPROM. EEPROM (electrically erasable PROM) and EAROM (electrically alterable ROM) are further variations, and these may be reprogrammed *in situ* by the system itself, although not normally as easily as RAM may be programmed. Such devices are useful in systems where a semi-permanent form of storage is required, but where the data may need to be modified on occasions.

RAM is capable of being both written to and read from. It is a very fast form of temporary store, in that data stored in it can be accessed typically in under 100 nanoseconds, but the store loses its data when the power is turned off (unless some form of battery backup has been arranged). Desktop computers often use a form of RAM called dynamic RAM (DRAM) whereby the memory devices are arranged on small plug-in modules called SIMMs (structured in-line memory modules), capable of storing many megabytes of data each. The amount of RAM available on a single chip increases every year, and thus it is difficult to be precise about storage capacity. RAM is used in computer systems for storing the temporary data which results from inputs and program operations, but also may be used for operating system software which has been temporarily uploaded from a more permanent external store such as a disk drive. It is important to distinguish between the RAM in a computer and its disk storage capacity, since they are both often referred to as 'memory'.

1.6.4 I/O ports

The I/O ports are the third main element of a basic microcomputer system. As mentioned earlier, these are the system's windows on the outside world, without which the piece of equipment would be virtually useless. Some of these ports may appear as physical connectors on the rear panel of the equipment concerned, and others may be connected to internal peripherals such as disk drives, in which case the ports may not be available for general purpose use. One very common use of I/O ports in some desktop computers is to route data to a video card, so that images can be displayed. I/O ports may be either parallel or serial, and this topic is discussed in more detail in section 1.8.5.

The important point to raise here is the way in which information is communicated to and from the ports, because there are a variety of ways of doing this, some of which will be outlined here. First of all, it will be seen from the main block diagram that the I/O ports in a simple computer are generally connected to the same data and address bus structure as the memory, but in some systems there is a separate control line from the CPU to indicate whether it is addressing memory or I/O – they are not both enabled at the same time. An alternative is to use so-called 'memory-mapped I/O', in which I/O ports occupy memory addresses. Each I/O port has at least one address, and usually a number, corresponding to different aspects of the port's function. One such address may be the port's 'control register' which allows the CPU to program such aspects as the clock speed of a serial interface. By writing to an I/O address the CPU can transfer a byte of data from a temporary internal store to a peripheral device.

Similarly, by reading a particular control register address it is possible to tell whether a byte of data has been received by a port. This is called 'polling', and is the CPU's means of determining whether there is any data to be had from any of the ports. It is rather like a teacher asking each member of the class in turn whether they want to say something – eventually the teacher may come across one who does, but if only two out of thirty have something to say it is a fairly time consuming and wasteful exercise. An alternative is for the person who has something to say to raise a hand and wait for the teacher to respond. In computer terms this is called 'interrupting', and is often used as a means of flagging-up the presence of new data at a port.

When a port receives a byte of data it may be able to raise the interrupt flag, which is one of the control lines connected to the CPU. The CPU then finishes the current task and acknowledges the interrupt using another line (IACK), after which a number of possible scenarios may occur. Either the CPU must then address each port in turn to find out which one interrupted, or, more efficiently, on receipt of the IACK flag, the port which generated the interrupt may place its own address or another unique identifier onto the data bus, which may then be read by the CPU to determine the source of the interrupt. This data value can be used to cause the CPU to jump its program counter to a new memory address which contains the start of a small subroutine to handle that particular interrupt, which in a simple case might involve reading the byte of data from the input buffer of the port and storing it in a RAM location. Such a process is one way of handling the arrival of MIDI commands at a serial input port.

When a lot of interrupts are likely to occur, such as when there are a large number of ports, it may be necessary to prioritise them, and often a dedicated device is used for this purpose which is capable of taking in a large number of interrupts and arranging for the CPU to service those with the highest priority first. Those which are time critical, such as MIDI synchronisation messages, will clearly have to be serviced before those which are not. Ports may need a small amount of buffer memory associated with them in order to allow data to be stored temporarily, in case the CPU is not able to service an interrupt immediately. If another byte of data were to arrive at the same port before the first one had been collected it would be lost unless there were some temporary storage.

Commonly encountered general purpose I/O ports on desktop computers are serial ports conforming to RS232 or RS422 standards, parallel ports such as the Centronics printer interface, and fast parallel ports such as SCSI (the Small Computer Systems Interface), which is often used for connecting hard disk drives.

1.6.5 Factors affecting computer system performance

Aside from the question of software design, there are a number of hardware issues that affect the performance of any system based around a microprocessor. It is useful to know about these because it may help when deciding what equipment to use for a certain purpose, and whether one device will perform better than another.

The main thing that most people are interested in when comparing computers is the speed with which they will perform certain operations. This is important because a faster computer will be capable of performing complex tasks without requiring the user to go away and make a cup of coffee while the computer works on the problem. A fast computer may be able to perform certain real-time operations that a slower machine might not, because real-time tasks such as animation, video, digital audio recording and music sequencing all require that a certain number of operations are completed in a specific time frame.

Clock speed is one of the factors dictating performance, since, as already explained, this dictates the rate at which instructions are sequenced. Early computers used in MIDI equipment tended only to be capable of clock speeds in the region of 2 MHz, which is slow by modern standards. An older desktop computer found today might be running at around 8 MHz, and the super-fast machines at rates up to around 80 MHz. Clock speeds of CPUs are increasing all the time, and so one should expect continued improvements in this area. It is not possible to say what minimum clock speed is required for a particular task, since it is only one of the determining factors.

Bus width is a second factor. In general, the wider the bus the faster the machine, since a wide bus makes it possible to transfer more data between devices per instruction. For example, a 4 byte floating point number (that is one with a number of decimal places and an exponent, or power of ten) would take four fetches to load it into the CPU with an 8 bit data bus, but only one with a 32 bit bus. It is very important, though, to distinguish between the width of the *system's* data bus and that within the CPU. The width of the CPU's internal data bus dictates what it can manipulate and store internally, and the external bus dictates how much of this data can be fed to and from the peripherals and memory in one go. Some so-called 32 bit CPUs have only 16 bit external data buses, whereas full 32 bit CPUs also have 32 bit external buses.

The speed of RAM installed in a system may also affect the apparent speed of operation, since slower RAM requires that the CPU waits a certain number of machine cycles for it to produce stable data after having been addressed (known as 'wait states'). Faster RAM allows the use of no wait states, thus speeding up access to stored data during program execution.

A particular type of processor, known as the RISC processor, is increasingly used in digital equipment because it operates using a reduced (simplified) instruction set in order to be able to run at very high speeds. For a given clock speed the RISC-based machine will tend to carry out more operations per second than a conventional microprocessor-based machine.

There are also a number of miscellaneous factors which will affect the performance of a computer-based device. These include whether or not any co-processors are installed, and whether any form of fast memory cacheing is used. Co-processors are additional CPUs designed to share or remove some of the processing load, and these often deal with such functions as mathematical operations or graphics processing. Whether a co-processor will give any improvement in speed depends largely on the task in hand, and whether the software is designed to benefit from co-processing. The same is true of fast memory cacheing, which is a means of storing the most recently used data in a small amount of very high speed RAM, either close to or installed in the CPU, such that it can be accessed more quickly and easily than general purpose RAM.

1.7 Mass storage

ROM and RAM are both solid-state forms of storage, and their advantage is that they can be built into the electrical bus structure of the computer and accessed very quickly, but there is often also the need for storage which can be removed or which can be used to keep much larger amounts of data than could reasonably or economically be stored in a solid-state form. The larger, more permanent forms of storage are called mass storage devices, and usually take the form of disk drives or tape drives. Using such devices it is possible to store an almost unlimited amount of data, especially if removable media are used. The access time to data stored on such peripherals tends to be quite a lot longer than for internal RAM, being of the order of a few milliseconds for the typical hard disk drive, and a number of seconds for tape drives.

Data stored on mass storage devices is formed into 'files', and a catalogue of file locations is kept in an index known as a 'directory'. Mass storage is often connected to computer I/O ports using the SCSI (Small Computer Systems Interface), which is a fast parallel interface commonly found on desktop computers and also often on MIDI-controlled samplers. Although there is not space here to cover the features of mass storage devices in detail, a summary will be given.

1.7.1 Winchester hard disks

The so-called 'hard disk' used in many computer systems is properly called a 'Winchester disk drive'. It is a sealed unit which contains a number of removable disk surfaces which are inflexible and mounted concentrically on a common spindle which rotates at high speed during operation. A diagram is shown in Chapter 5. The disks are coated with a magnetic material, and the recording process is magnetic. Each surface has a read/write head which can be moved quickly across the surface using a positioner arm, in order to access or store data in one of the predefined locations – the head flying a small distance above the disk's surface owing to the lift provided by the rotating disk. The disk surfaces are divided up, during a process known as 'formatting', into a number of sectors each with a unique address (as shown in Figure 5.30), and into a number of concentric rings known as cylinders. On an individual surface the sector becomes a 'block' and the cylinder a 'track'. A block may contain a fixed number of bytes, often 512 or 1024 bytes.

The disks in a Winchester drive are not removable, but the drive has the advantage of being extremely fast to access and transfer data (although slower than RAM). Typically a block may be accessed in around 5–10 ms, and improvements are regularly being made in this respect. The storage capacity of a Winchester is subject to regular increases as technology improves, and currently may be anywhere from around 40 Mbytes to over 3 Gbytes. The transfer rate of data to or from a Winchester is often limited by the speed with which a computer or interface can accept it, and typical SCSI interfaces vary between about 1 and 4 Mbytes/s depending on the computer. There is also a SCSI-2 interface which is faster than ordinary SCSI, and can be used for transfer between very fast disk drives and computers.

Winchester disks are sometimes used with MIDI-controlled music samplers as a means of more permanent storage for large quantities of audio sample data which may be accessed relatively quickly. A 330 Mbyte hard disk, for example, is capable of storing roughly an hour of full bandwidth digital audio.

1.7.2 Floppy disks

Floppy disks are low speed, low capacity disks made of flexible material, coated with a magnetic oxide. Like hard disks they are formatted into individually addressable tracks and blocks. The read/write head of the drive is in contact with the disk surface during operation, and can only move relatively slowly – the access time being of the order of hundreds of milliseconds. Floppy disks are used where removability is important, although they have a relatively low capacity when compared with hard disks, storing a maximum of only a few megabytes.

A relatively recent development exists in the form of the 'floptical' disk, which is a floppy disk capable of holding some 20 Mbytes of data in a compact form by taking advantage of a combination of optical and magnetic processes. In this system the data storage is magnetic but the positioning of the read/write head is controlled optically.

1.7.3 Optical disks

A number of types of optical disks exist, and it is not proposed to describe them all here, but it will be sufficient for the purposes of this book to say that the general principle of optical disks is that they are written to and read from using a low-powered laser whose light is reflected off the disk's surface. By modifying the disk's surface in one of a variety of ways, its reflectivity may be varied during the writing (recording) process so as to differentiate between binary data states. The great advantage of optical disks over Winchesters is their removability, coupled with moderately high capacity. They are slightly slower in operation than Winchester drives as a rule.

The track on an optical disks is normally a continuous spiral rather than concentric rings, but the formatting process will aim to divide this up into conventional sectors for data storage purposes. Disks fall into two categories –

constant angular velocity (CAV) recording and constant linear velocity (CLV). In CAV recording the rotational speed of the disk remains constant no matter where the head is, whereas in CLV recording the disk rotates faster when the head is near the centre compared with at the edge, maintaining a constant linear speed of the track under the head. Other methods such as ZCAV (zoned CAV) recording are sometimes used to store a greater amount of data in outer tracks on CAV disks, requiring more flexible clocking and control mechanisms in the disk drive.

Optical disks may be read only, write-once–read-many-times (WORM), or fully erasable and rewritable. Magneto-optical (M-O) principles are now used widely in the manufacture of erasable disks, being based on a combination of magnetic and optical recording. The writing process involves heating a spot on the disk using a laser, and then altering its magnetic polarisation using a magnet. Owing to the special properties of the rare earth recording layer, the magnetic polarisation of a spot influences the polarisation of reflected light on replay, allowing the data states to be detected during the read process by using a purely optical process which detects small changes in reflected light polarisation. The recorded state may be reversed many times by re-heating the recording layer and reversing the magnetic state. There are currently two ISO standard formats for M-O disks, one for 3.5-inch disks storing 128 Mbytes, and one for 5.25-inch disks storing around 600 Mbytes, but some manufacturers have extended the capacity of 5.25-inch disks to nearly 1 Gbyte using non-standard modes of operation. There are also new optical disk formats appearing, capable of storing over 1 Gbyte.

The compact disc (CD) is used widely in data storage, as well as in audio, and there are many formats. The CD-ROM is a conventional CD that has been formatted into blocks and given a data structure rather like a hard disk so that it can be used to store non-audio data. It is possible to store over 600 Mbytes of data on a CD in a read-only form, making it an ideal medium capacity store for permanent data which is to be widely and cheaply distributed. There is also a write-once version of the CD available, at moderate cost (currently about £20), which can be used to store either audio data or computer data, and erasable versions are feasible although not yet available. The main limitation with the CD is its access time, since the normal rotational speed of the CD is between 200 and 500 rpm using CLV recording, making the access time of the order of hundreds of milliseconds (although double-speed and quad-speed CD-ROM drives are now beginning to appear on the market).

1.7.4 Miscellaneous mass storage media

Various types of tape cartridge drive are used for cheap storage of large volumes of data where access time is not a priority. These include the Exabyte system which is based on a Video-8 tape, the QIC (quarter-inch cartridge) and the DDS system which is based on a DAT tape. Like the various disk formats, the tape is formatted into addressable blocks and a directory is stored somewhere on the tape to index the contents.

There also exist a number of removable magnetic disks, such as the Bernoulli and SyQuest mechanisms, which are designed to store between roughly 40 and 120 Mbytes of data on a 5.25-inch removable cartridge. These are reasonably fast (faster than a floppy and approaching the speed of a Winchester), and can be used where a floppy is not large enough for the files required. One should expect to see increased-capacity products of this type in the near future.

1.7.5 Disk and tape formats

A practical problem arises when attempting to transfer data from one system to another using disks (or tapes for that matter). Different systems format the data in different ways, such that the block and directory structure are not the same, making it more difficult to exchange data. Even if the physical disk format will fit into the disk drive of another machine, there is no guarantee that the drive will be able to read the data stored on it, or that the system will interpret the files concerned. There is a trend at the moment towards making multi-standard hardware and software capable of reading a variety of disk formats, but the situation is far from perfect. Anyone intending to transfer data between dissimilar systems should investigate the individual situation carefully to determine the likelihood of success. Various software utilities exist to convert files from one format to another, and some software applications use 'filters' or 'translators' to enable them to interpret foreign file formats.

1.8 Digital communications

MIDI is a means of transferring digital information from one device to another, and as such it falls within the larger field of digital or data communications. In this section a number of fundamental aspects of this field relevant to MIDI will be introduced in order that the reader may be familiar with some of the most important terminology and techniques. It is also increasingly common for MIDI systems to involve networks of one kind or another, and therefore an introduction will be given to the principles of network operation. Although digital interfaces such as MIDI are primarily concerned with the transfer of binary data, there are 'analogue' problems to be considered, since the electrical characteristics of the interface such as the type of cable used, its frequency response and impedance will affect the ability of the interface to carry data signals over distances without distortion.

1.8.1 Wire and fibre connections compared

Considering only physical interconnects, it is possible to carry data over both electrical wire and optical fibre. Wire has the advantage that it is relatively easy to deal with, and it can be interfaced simply. Signals travelling down wire can suffer considerable distortion and attenuation (loss) as distance increases. Fibre, on the other hand, is more resilient to noise and interference than wire, and has a very wide bandwidth (that is it can carry a large amount of information at high speed). Fibre can also carry data over very long distances with minimal attenuation of the signal, and thus it is used for such purposes as transatlantic telephone links and large, high speed computer networks. Also, when fibres are used as interconnects, devices are electrically isolated from each other.

1.8.2 Balanced and unbalanced electrical interfaces compared

In an unbalanced interface there is one signal wire and a ground, and the data signal alternates between a positive and negative voltage with respect to ground (see Figure 1.16). Alternatively a current loop can be used, as with MIDI, in which the current is turned on and off according to the data state. The shield of the cable is

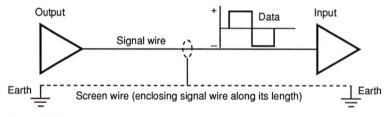

Figure 1.16 Electrical configuration of an unbalanced interface

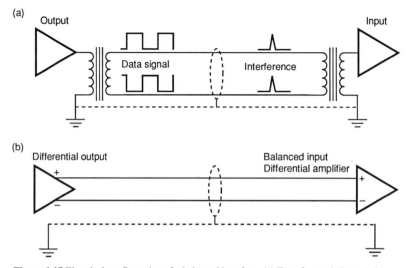

Figure 1.17 Electrical configuration of a balanced interface. (a) Transformer balanced. (b) Electronically balanced

normally connected to the ground at the transmitter end, and may or may not be connected at the receiver end depending on whether there is a problem with earth loops (a situation in which the earths of the two devices are at different potentials, causing a current to circulate between them, sometimes resulting in hum induction into the signal wire). The unbalanced interface tends to be quite susceptible to interference, since any unwanted signal induced in the data wire will be inseparable from the wanted signal. MIDI is an unbalanced interface.

In a balanced interface there are two signal wires and a ground (see Figure 1.17) and the interface is terminated at both ends either in a differential amplifier or a transformer. The driver drives the two legs of the line in opposite phase, and the advantage of the balanced interface is that any interfering signal is induced equally into the two legs, in phase. At the receiver any interference is cancelled out either in the transformer or differential amplifier, since such devices are only interested in the difference between the two legs. The pair of wires carrying the data signal is often twisted together – the so-called 'twisted pair' – thus ensuring that they both receive roughly equal exposure to interference, and to ensure that the spacing between the conductors remains constant on average.

The balanced interface therefore requires one more wire than the unbalanced interface, and will usually have lines labelled 'Ground', 'Data+' and 'Data–', whereas the unbalanced interface will simply have 'Ground' and 'Data'. For temporary test setups it is sometimes possible to interconnect between balanced and unbalanced electrical interfaces or vice versa, by connecting the unbalanced interface between the two legs of the balanced one, or between the ground and one leg, but often the voltages involved are different, and one must take care to ensure that the two data streams are compatible. The RS422 interface found on some desktop computers is an example of a balanced interface.

1.8.3 Cable effects

At low data rates (MIDI uses a fairly low data rate) a piece of wire can be considered as a simple entity which conducts current and perhaps attenuates the signal to some extent, and in which all components of the signal travel at the same speed as each other. The cable will exhibit a certain amount of capacitance between the conductors (due to their separation) and as the frequency of the signal rises this capacitance will become increasingly important. Cable inductance may also become significant in long interconnects. The resistance to current flow is determined by the resistance of the cable (affected by its cross-sectional area), and the longer the cable the greater this resistance – resulting in a reduced overall signal amplitude at the receiving end. If a cable is moderately short its capacitance appears lumped as a discrete component, lying between the signal wire and the ground, thus forming a link between the two for signal to leak across. The effect, as shown in Figure 1.18, is to roll off the otherwise sharp edges of the data signal, leading to rise-time distortion. This affects the timing of the signal, and in extreme cases may render it unreadable by a receiver. This is the effect noticed when particularly long MIDI cables are used, where data becomes corrupted due to the excessive losses in the cable, although the data rate of MIDI is only fairly modest at 31.25 kbit/s.

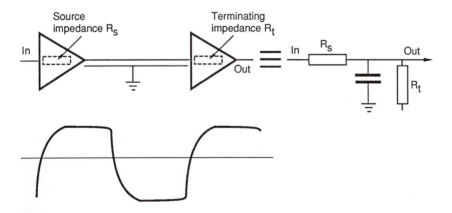

Figure 1.18 Cable capacitance rolls off the edges of the data signal leading to rise-time distortion

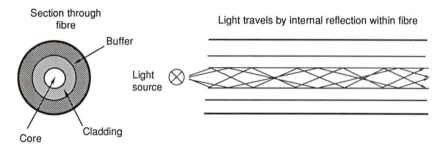

Figure 1.19 Cross-section through an optical fibre, showing how light travels

1.8.4 Optical fibres

Data is transferred over fibres by modulating a light source with the data, the light being carried within the fibre to an optical detector at the receiving end. The light source may be an LED or a laser diode, with the LED capable of operation up to a maximum of a few hundred megahertz at low power, whilst the laser is preferable in applications at higher frequencies or over longer distances.

The typical construction of an optical fibre is shown in Figure 1.19, and light travels in the fibre as in a typical 'waveguide', by reflection at the boundaries. Total internal reflection occurs at the boundaries between the fibre core and its cladding, provided that the angle at which the light wave hits the boundary is shallower than a certain value called the critical angle, and thus the method of coupling the light source to the fibre is important in ensuring that light is propagated in the right way. Unless the fibre is exceptionally narrow, with a diameter approaching the wavelength of the light travelling along it (say between 4 and 10 μm), the light will travel along a number of paths – known as 'multimodes' – and thus the time taken for a source pulse to arrive at the receiver may vary slightly depending on the length of each path. This results in smearing of pulses in the time domain, and is known as modal dispersion, or, looked at in the frequency domain, it represents a reduction in the bandwidth of the link with increasing length. Losses are usually quoted in dB per km at a specified wavelength of light, and can be as low as 1 dB/km with high quality silica, graded index multimode fibres, or higher than 100 dB/km with plastic or ordinary glass cores.

Single mode fibres with very fine cores achieve very wide bandwidths with very low losses, and thus are suitable for use over long distances. Attenuations of around 0.5 dB/km are not uncommon with such fibres, which have only recently become feasible due to the development of suitable sources and connectors.

1.8.5 Serial and parallel interconnects

The bits of a binary word may be transmitted either in parallel or serial form. In the parallel form each bit is carried over a separate communications channel and the result is at least as many channels as there are bits in the word. Thus a 24 bit parallel interface would require at least 24 wires, an earth return, a clock line, and a number of address and control lines (there are normally additional lines for controlling the exchange of data – called 'handshaking', as described in section 1.8.9). Parallel interconnection is normally used for short distance communication paths ('buses') within a computer-based device because it can be extremely fast, but is bulky and uneconomical for use over longer distances.

When data is carried serially it only requires a single channel (although electrically that channel may consist of more than one wire), and this makes it economical and simple to implement over large distances. On a serial interface the bits of a word are sent one after the other, and thus it tends to be slower than the parallel equivalent, but the two are so different that to say this is really oversimplifying the matter since there are some extremely fast serial interfaces around. Data is converted from the parallel form, usually used within computer equipment, to the serial form using a shift register as explained in section 1.3. The rate at which the clock runs determines the data rate of the interface. Some serial interfaces carry clock and control information over the same channel as the data, whereas others accompany the data channel with a number of additional parallel lines and a clock signal to control the flow of data between devices. Depending on the protocol in use it is possible for serial data to be sent either MSB first or LSB first, and knowing the convention is clearly important when interpreting received data. The MIDI interface is a serial interconnect.

Many serial interfaces are engineered using an I/O device known as a UART (universal asynchronous receiver/transmitter), which is a sophisticated form of shift register. A UART has both input and output, and can convert serial to parallel data and vice versa simultaneously. It has a clock input, or may generate its own clock, and the frequency of this can often be programmed by writing to a control register which resides at a different address to the I/O port itself. The UART is able to format the data to be transmitted, and can synchronise reception to incoming data. It will raise a flag in its control register when it has received a byte of data, and can store that byte temporarily until it is collected.

Interface standards specify various peak-to-peak voltages for the data signal, and also specify a minimum acceptable voltage at the receiver to ensure correct decoding (this is necessary because the signal may have been attenuated after passing over a length of cable). Quite commonly serial interfaces conform to one of the international standard conventions which describe the electro-mechanical characteristics of data interfaces, such as RS422 which is a standard for balanced communication over long lines devised by the EIA (Electronics Industries Association), or RS232 which is an unbalanced interface with a number of auxiliary control lines used widely for connecting telecommunications equipment such as modems. (Standards such as RS422 are mostly only electrical or electro-mechanical standards, and do not say anything about the format or protocol of the data to be carried over them.)

1.8.6 Data rate versus baud rate

The data rate of an interface is the rate at which information is transferred (sometimes referred to as the information rate), whereas the 'baud rate' of an interface is the modulation rate or number of data 'symbols' per second. Although in many cases the two are equivalent, modulation schemes exist which allow for more than one bit to be carried per baud.

Data rate is normally quoted as so many kilo or megabits per second (kbit/s, Mbit/s) and this must normally include any capacity for control and additional data. MIDI carries one bit per baud, and therefore its data rate of 31.25 kbit/s corresponds to 31.25 kbaud.

1.8.7 Synchronous and asynchronous communications

The receiving device must be able to determine the time slot in which it should register each bit of data which arrives. In synchronous communications a clock signal normally accompanies the data, either on a separate wire or modulated with the data, and this is used to synchronise the receiver to the transmitted data. Each bit of data may be registered at the receiver on one of the edges of the separate clock, or the clock may be extracted from the modulated data using a process known as clock recovery.

In asynchronous communication the clocks of the transmitter and receiver are not locked directly, but must have an almost identical frequency. The clock rate tolerance is often around ±1%. In such a protocol each byte of data is prefixed with a start bit and followed by one or more stop bits (as shown in Figure 2.3) and the phase of the receiver's clock is adjusted at the trailing edge of the start bit. The following data bits are then clocked in with reference to the receiver's clock, which should remain sufficiently in phase with the transmitted data over the duration of one byte to ensure correct reception. The receiver's clock is then resynchronised at the start of the next data byte. These functions can be performed by a UART, as described above. Such an approach is often used in computer systems for exchanging data with remote locations over a modem, for example, where the gaps between received bytes may be variable and data flow may not be regular. MIDI is an asynchronous interconnect.

1.8.8 Uni- and bidirectional interfaces

In a unidirectional interface data may only be transmitted in one direction, and no return path is provided from the receiver back to the transmitter. In a bidirectional interface a return path is provided and this allows for two-way communications. The return path is often used in a simple serial situation to send back handshaking information to the transmitter, telling it whether or not the data was received satisfactorily.

A simplex interface is one which operates in one direction only; a half-duplex interface is one which operates in both directions, but only one at a time; and a full-duplex interface is one capable of simultaneous transmission and reception. MIDI is a unidirectional interface, but two-way communications may be set up using a pair of MIDI interconnects.

1.8.9 Controlling data flow

As mentioned above, a process known as handshaking is often used for controlling the flow of data between two devices. The alternative is 'blind' transmission, in the hope that the receiving device will receive the information, but without really knowing.

In a controlled data transfer where time is not critical, it is possible for a transmitter to indicate that it is ready to send some data. The receiver may then respond that it is ready, and may acknowledge the correct receipt of a packet of data. If the packet has been received with an error then the receiver may request retransmission. This is particularly useful when dealing with noisy interconnects where data transfer may be unreliable, or when the receiver is expected to be busy a lot of the time (in which case data could be lost if transmission began without a go-ahead). In serial interfaces it is sometimes possible for this exchange of pleasantries to take place over the same lines as the data transfer, with a return link from the receiver being used to signal correct reception to the transmitter (this is so-called 'software handshaking'). Alternatively, if the devices are relatively close together, additional lines may be used between the devices dedicated to the handshaking process – called 'hardware handshaking'.

In real-time applications such as MIDI control it may not be possible for such protocols to be observed, since an erroneous byte intended to control an event would not really be any use if retransmitted later than its appointed time. It is therefore necessary to ensure that a receiver is capable of dealing with data as it arrives, having a short-term buffer (temporary store) to hold any data which might arrive during a time when the receiver is unable to respond immediately. Information which is time critical must be dealt with at a higher priority than that which is not.

Certain non-realtime applications of MIDI, such as the audio sample dump (see section 3.14.2) allow a return link to be set up so as to allow for a controlled dump of large amounts of data, since the buffer sizes, speeds and memory capacities of receiving devices vary considerably, and there would be a possibility of lost data without handshaking.

1.9 Networks

A network may be considered as an extension of the simple point-to-point interfacing concept. It is a communications bus, nearly always serial for ease of installation, which allows data to be exchanged between any of a number of devices, all of which are connected to the network (see Figure 7.14). In a simple network, any data transmitted by one device is detected by all devices, and thus some means is required of identifying the data intended for each device. In order to manage the flow of data and to allow a device to determine the destination of a transmitted data packet, each device on the network is given an address (set either in hardware or software). Data is transmitted in packets (groups of bytes) which are preceded by a header denoting the destination, amongst other things.

A serial network normally runs at a higher data rate than a simple point-to-point interface, since it is to be 'time-shared' or multiplexed between a number of users, each of which may require a reasonable data rate. The network control hardware and software for each device looks after the job of finding an empty slot on the network, and may be designed so that no one device can 'hog' the network to the exclusion of others. Such an approach of management is normally only necessary in large networks where the capacity may be approaching maximum usage on occasions. The level of usage of a network determines the speed required, and too slow a speed will result in delays since there is less likelihood that a device will be able to transmit a packet when it wishes to. It is rather like a motorway: when the motorway is empty, cars entering it from side roads can do so almost whenever they arrive, with no delay, but when the motorway is full queues may build up on the side roads while the cars wait to enter the motorway. If the speed of cars on the motorway is increased, or if more lanes are added, it can handle more cars per second, and thus the queues on the side roads will diminish. Just as with networks, there is no point in having a four-lane motorway if you are only intending to run a few cars on it.

Networks tend to be divided up into categories depending on their size. A local area network (LAN) will normally operate within a single building, perhaps with various 'zones' designed to limit network traffic to particular work areas, with controlled 'bridges' between the zones. A metropolitan area network (MAN) will cover a wider area such as a town, and a wide area network (WAN) could be, say, national. Bridges may be provided at each junction between two networks, allowing traffic to pass in each direction. A simple LAN such as Apple Computer's LocalTalk runs at a slow rate of roughly 0.25 Mbit/s – designed to handle traffic from a small number of devices generating limited amounts of data, whereas a medium speed LAN such as Ethernet runs at 10 Mbit/s. Fast networks based on optical fibres, such as the FDDI (Fibre Distributed Digital Interface), operate at speeds of around 100 Mbit/s.

As multimedia uses of computer systems increase, requiring the real-time transfer of digital audio, video and control data (such as MIDI) over networks, so the need for increased network speed grows. Network speeds which once were acceptable for transferring text files from place to place are now wholly inadequate for these new applications. Communication protocols now exist specifically designed for real-time applications, and these will be used increasingly in preference to older protocols. A real-time network protocol will need to guarantee certain limits on the delay of data packets, and may need to allow a proportion of the network bandwidth to be reserved for communication between two devices for a given period of time. A small number of network products have been developed specifically for studio and live control applications, which allow a selection of allow audio, video and MIDI information, along with other data if required, to be routed between multiple devices over optical fibres. These allow multiple MIDI data streams to be multiplexed on the same fibre, running at sufficient speed to make delays unnoticeable. This is discussed further in section 7.8.

1.10 The role of software in computer systems

Software has been mentioned a number of times in the preceding sections of this chapter, and it will be mentioned many times again, but what is it and what is its role in the computer system? This section will give an introduction to the two main areas of software which most people will come across – operating systems and applications software. A number of so-called 'real-time operating systems' now exist for use with MIDI software, and these are designed as extensions to a computer's main operating system which optimise the handling of MIDI data, often for more than one application at a time, in order to minimise delays and ensure the fastest possible throughput of data (this is covered in more detail in section 5.5). An understanding of the nature of operating systems will help the person trying to understand MIDI applications appreciate how a particular software application interacts with the computer's more hidden workings, and how applications interact with each other.

1.10.1 What is software?

The CPU of a computer is a logical machine which can step through a preprogrammed sequence of operations at a speed defined by the clock. It has no magical 'knowledge' or 'skill' and it relies on being told what to do next. Depending on data presented to the CPU at different stages in its machine cycle, the sequence of operations may be changed. This is really the role of software – to program the CPU and to provide it with some of the data to be operated upon by the program instructions (the rest comes from the outside world, or is the result of operations performed during the execution of the program).

Within the CPU is an instruction decoder (see section 1.6.2) which reads a byte of data (software) supplied from the memory and sets a complex collection of logical gates, latches and registers to a particular state, so that on the next cycle of the clock a particular collection of logical changes will occur. It is an electronic version of the mechanical adding machine where the user sets up certain mechanical conditions within a device consisting of cogs, ratchets and levers, after which the turning of a handle (like the cycling of the clock) executes a mechanical sequence which depends on the initial mechanical state of the components.

Thus software is a sequence of binary instructions and data which causes the CPU to act in a particular way as the clock cycles. Software at this level is called 'machine code', and the instructions are called 'low level' instructions, since they are at the raw binary level which the CPU can interpret directly. It is this data which is actually stored in the computer's memory and which is used by the CPU, but unfortunately to write a program at this level is exceedingly tedious.

What is required is a form of 'high level' language, and there are many of these with different strengths. One up from machine code is 'assembler' which is really a collection of mnemonics representing machine code instructions which are easier to use than pure binary codes. Writing a program in assembler is very tedious but quite efficient from a running point of view, because it is written at the 'nuts and bolts' level of the system. At the highest level are programming languages like 'Hypertalk' that look almost like speech, containing lines like:

put the contents of field 1 into line 3 of field 5

and these languages are designed for use by people who want to write their own software but do not wish to get too deeply into the inner workings of the computer. The drawback of many such languages is that the programs run more slowly because a lot of interpretation has to take place to turn these 'English' commands into machine code before the operation can be executed.

1.10.2 Compiled and interpreted programs

There are two basic approaches to turning high level programs into machine code – compiling and interpreting. A compiler is a software program which takes the whole of a high level program and turns it into efficient machine code before it can be run. Execution is then a matter of running the machine code or compiled version of the software. This approach is tedious when writing the program in the first place – because the software has to be compiled each time before it can be run – but it is used widely because the compiled program then runs very quickly. An interpreter codes the program lines into machine code *while* the program is running, and thus it is convenient to use while programming but very slow in execution due to the need for interpreter intervention all the time.

The two approaches are very similar to language translation for speech. If you talk to a foreign person via an interpreter it is quite convenient but you have to keep stopping every line for the interpreter to translate what you said. If what you wanted to say was written down on paper and translated (compiled) first, then either you or a native speaker could use it directly and more quickly, but it would require an initial delay for the translation.

1.10.3 Object-orientated software

Object-orientated software is a concept which has grown considerably in importance over the last few years, and it presents another level of high level interaction with the computer. In an object-orientated environment, software 'modules' are written, each of which performs a particular function, and these are termed software 'objects'. An object may have a number of inputs and outputs. The objects may then be combined and interconnected, often graphically, in various ways such that the output of one object feeds another's input, and so on, creating what is really a virtual machine with controls, processors, inputs, outputs and displays. The object-orientated concept extends into corresponding programming languages as well.

1.10.4 Operating systems

An operating system (OS) is a software program which runs 'in the background' all the time that a computer-controlled device is turned on. It is this which gives a particular computer system its peculiar characteristics, such as how it displays information, what commands it accepts, how disks are formatted and how memory is organised. It is also likely to be dedicated to a particular family of CPU, because the operating system must issue low level instructions in CPU-specific code. The operating system is a fundamental 'toolkit' which is called upon by higher level applications to perform the more mundane tasks such as disk storage, handling I/O from the keyboard, writing to the display, and so on. It gives a degree of consistency to the operation of a particular system, and avoids the need for programmers to write very basic functions such as those dealing with mathematical operations every time they write an application. The operating system is a layer of software intervention which resides in between the application and the microprocessor, as shown in Figure 1.20. It is still possible for applications to deal directly with the CPU, but most tasks can be passed via the operating system.

Examples of common personal computer operating systems are Microsoft's MS-DOS (designed for 80x86 CPU family) and Apple's System 7 (designed for the 680xx series of CPUs). Larger computers and mainframes often operate under the Unix system. Applications written for these operating systems must conform to certain basic guidelines which describe how the program should interact with the operating system. Older operating systems tended to be text based, in that the only

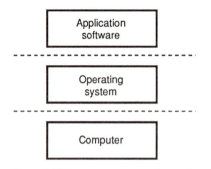

Figure 1.20 The operating system forms a layer between the application software program and the inner workings of the computer

way that a user could interact with the OS was by typing strings of often unrememberable commands from a QWERTY keyboard, every character and space of which had to be correct otherwise the user would be presented with an equally incomprehensible error message. Recent systems have been designed to be more 'user friendly', often employing graphical user interfaces, as discussed below.

1.10.5 Graphical user interfaces

Graphical user interfaces, or GUIs as they are sometimes known, do not use text as the main means of communication with the operating system. Instead a whole graphical 'layer' is placed between the user and the operating system such that the user may only have to point at what he wants and select it. Often a mouse or other type of pointing device is interfaced to the computer via a port, to control a displayed arrow on the screen which moves as the mouse moves. A button on the mouse is then used to select whatever is being pointed at. Instead of the user having to remember commands precisely, a choice of options is presented in menu form and the user selects whichever is required. Such environments have been called WIMP (Windows, Icons, Mouse and Pointer) environments.

GUIs have done much to bring the power of the desktop computer to those who don't feel 'computer-minded', although dedicated computer 'boffins' often profess a dislike for them because they tend to slow down the operating system. This effect on speed is true, but the benefits far outweigh the disadvantages, and the dislike of the GUI by such people is probably more to do with a wish to continue the pretence that computers are only for people who understand them!

A GUI such as 'Windows' was designed as a layer on top of an existing operating system (MS-DOS), in an attempt to make it more user friendly. The result is rather slower than a GUI which was designed as an integral feature of the OS from the start, such as Apple's System software. It is likely that GUIs of some sort will be a key feature of computer OS's for the foreseeable future, possibly coupled with elements of voice control and handwriting recognition as means of providing alternative commands.

1.10.6 Filing structures

In a text-based OS such as MS-DOS, files on a disk are grouped into 'directories', which are virtual catalogues of subsets of the disk contents. Directories are a useful way of subdividing the disk contents in order that a track can be kept of stored data, and projects can be grouped together. An almost infinite number of subdirectories is possible, dividing the disk contents in a type of tree structure, as shown in Figure 1.21. A GUI may present this tree structure as a collection of nested folders, within which can be files or more folders. The text-based OS will allow the user to define a 'path' to the required file by typing in a text string representing the directories concerned and the file name, e.g.:

c:songs\song3.mid

whereas a GUI will allow this to be done using a means of opening up folders or selecting the appropriate subdirectory from a menu.

File names may be limited to a certain number of characters, and may allow the addition of an extension, often of three letters, to define the file as of a particular

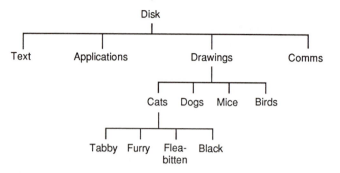

Figure 1.21 Disk files are organised in a tree-like structure, as shown in this example. Each directory can have subdirectories, down to many levels, within which may be files or further directories

type. They may also forbid the use of certain characters such as spaces. GUI OS's such as the Macintosh allow long filenames (up to 32 characters) with almost no restriction on what characters can be used, allowing files to be given more meaningful names such as '1993 Accounts (2nd draft)'.

1.10.7 Operating system extensions

An extension is a piece of software which adds to the OS's features or characteristics, and is often loaded from disk when the system starts up. It resides in memory and operates in the background, just as if it were part of the OS. Examples of extensions are the various MIDI OS extensions that are written by sequencer manufacturers, designed to optimise the performance of their software, and to tailor the computer's mode of operation to the handling of MIDI data. It is quite common for extensions to cause problems in computer systems, since they may interfere with the normal mode of operation and cause some applications to function incorrectly or even fail altogether. This is normally a matter for experimentation, and there are a number of utilities which exist to manage the turning on and off of system extensions.

1.10.8 Multitasking

A true multitasking OS is one which allows a number of applications to reside in different areas of RAM and run concurrently. The number of applications which may do this is normally limited by the amount of RAM, although a technique known as 'virtual memory' makes it possible to treat a proportion of disk capacity as if it were (slower) RAM. Multitasking can be useful when, for example, one application is performing a tedious task and the user wishes to work on another while the tedious task progresses. MIDI operating systems may attempt to run a number of applications concurrently, all synchronised to the same clock source, each handling a different aspect of studio operation. Multitasking requires a fast computer because it is required to perform more than one job at a time. When heavy processing tasks are taking place in the background, the foreground application can run much more slowly, and on slower machines this can make the foreground application virtually unusable.

A real-time multitasking system is one in which time is shared between applications in such a way that actions are completed within certain defined time limits. A number of PC operating systems are not truly multitasking in this respect. Multitasking should be distinguished from the simpler 'program switching' arrangement in which a number of applications reside concurrently in RAM and may be switched between. With program switching, inactive applications do not normally continue to operate in the background, although it is still a useful thing to have since it saves closing one job and loading another application from disk every time you wish to change between them.

Chapter 2

Introduction to MIDI control

In this chapter the principles of MIDI will be discussed, both in hardware terms and in terms of its function as a serial communications interface between devices. Many of the terms used in this chapter and in the remainder of the book have been introduced in Chapter 1, and thus will not be elaborated upon here. The latter part of the chapter introduces the concepts of simple MIDI control, and outlines the communications protocol used between devices.

2.1 Before MIDI

Electronic musical instruments existed widely before MIDI was developed in the early 1980s, but no universal means existed of controlling them remotely. Many older musical instruments used analogue voltage control, rather than being control-led by a microprocessor, and thus used a variety of analogue remote interfaces (if indeed any facility of this kind was provided at all). Such interfaces commonly took the form of one port for timing information, such as might be required by a sequencer or drum machine, and another for pitch and key triggering information, as shown in Figure 2.1. The latter, commonly referred to as 'CV and gate', consisted of a DC

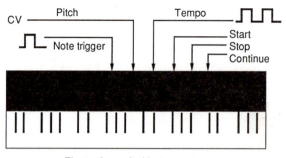

Electronic musical instrument

Figure 2.1 Prior to MIDI control, electronic musical instruments tended to use a DC remote interface for pitch and note triggering. A second interface handled a clock signal to control tempo and trigger pulses to control the execution of a stored sequence

(direct current) control line carrying a variable control voltage (CV) which was proportional to the pitch of the note, and a separate line to carry a trigger pulse. A common increment for the CV was 1 volt per octave (although this was by no means the only approach), and notes on a synthesiser could be triggered remotely by setting the CV to the correct pitch and sending a 'note on' trigger pulse which would initiate a new cycle of the synthesiser's envelope generator (see section 3.1). Such an interface would deal with only one note at a time, but many older synths were only monophonic in any case (that is, they were only capable of generating a single voice).

Instruments with onboard sequencers (see section 3.1.4) would need a timing reference in order that they could be run in synchronisation with other such devices, and this commonly took the form of a square pulse train at a rate related to the current musical tempo, often connected to the device using a DIN-type connector, along with trigger lines for starting and stopping a sequence's execution. There was no universal agreement over the rate of this external clock, and frequencies measured in pulses per musical quarter note (ppqn), such as 24 ppqn and 48 ppqn, were used by different manufacturers. A number of conversion boxes were available which divided or multiplied clock signals in order that devices from different manufacturers could be made to work together.

As microprocessor control began to be more widely used in musical instruments a number of incompatible digital control interfaces sprang up, promoted by the large synthesiser manufacturers, some serial and some parallel, and needless to say the plethora of non-standardised approaches to remote control made it difficult to construct an integrated system, especially when mixing equipment from different manufacturers. Owing to collaboration between the major parties in America and Japan, the way became cleared for agreement over a common hardware interface and command protocol, resulting in the specification of the MIDI standard in late 1982/early 1983. This interface grew out of an amalgamation of a proposed universal interface called USI (the Universal Synthesiser Interface) which was intended mainly for note on and off commands, and a Japanese specification which was rather more complex and which proposed an extensive protocol to cover other operations as well. Since MIDI's introduction, the use of older remote interfaces has died away very quickly, but there remain available a number of specialised interfaces which may be used to interconnect non-MIDI equipment to MIDI systems by converting the digital MIDI commands into the type of analogue information described above.

In the last ten years the standard has been subject to a number of addendums, extending the functionality of MIDI far beyond the original. The original specification was called the MIDI 1.0 specification, and since it have been such additions as the MIDI Sample Dump protocol, MIDI Files, General MIDI, MIDI TimeCode, MIDI Show Control and MIDI Machine Control, all of which are discussed in this book. A number of organisations exist to coordinate and promote changes and additions to the MIDI standard, these being principally the International MIDI Association (IMA), the MIDI Manufacturer's Association (MMA) and the Japanese MIDI Standards Committee (JMSC). The latter two are manufacturer's associations which act to coordinate those aspects of the standard which relate directly to implementation in equipment, such as the definition of previously undefined options and manufacturer identifications in system exclusive messages. Contact information for the IMA is provided in Appendix 1.

2.2 What is MIDI?

MIDI is the Musical Instrument Digital Interface, and, as its name suggests, it is a digital remote control interface for musical systems. It follows that MIDI-controlled equipment is normally based on microprocessor control, with the MIDI interface forming an I/O port (see section 1.6). It is a measure of the popularity of MIDI as a means of control that it has now been adopted in many other audio and visual systems, including the automation of mixing consoles, the control of studio outboard equipment, the control of lighting equipment and of other studio machinery. Although many of its standard commands are music related, it is possible either to adapt music commands to non-musical purposes or to use command sequences designed especially for alternative methods of control.

The adoption of a serial standard for MIDI was dictated largely by economic and practical considerations, as it was intended that it should be possible for the interface to be installed on relatively cheap items of equipment, and that it should be available to as wide a range of users as possible. A parallel system might have been more professionally satisfactory, but would have involved a considerable manufacturing cost overhead per MIDI device, as well as parallel cabling between devices, which would have been more expensive and bulky than serial interconnection. The simplicity and ease of installation of MIDI systems has been largely responsible for its rapid proliferation as an international standard.

Unlike its analogue predecessors, MIDI integrates timing and system control commands with pitch and note triggering commands, such that everything may be carried in the same format over the same piece of wire. MIDI makes it possible to control musical instruments polyphonically in pseudo real time: that is, the speed of transmission is such that delays in the transfer of performance commands are not audible in the majority of cases. It is also possible to address a number of separate receiving devices within a single MIDI data stream, and this allows a controlling device to determine the destination of a command.

2.3 MIDI and digital audio contrasted

For many the distinction between MIDI and digital audio may be a clear one, but those new to the subject often confuse the two. Any confusion is often due to both MIDI and digital audio equipment appearing to perform the same task – that is the recording of multiple channels of music using digital equipment – and is not helped by the way in which some manufacturers refer to MIDI sequencing as digital recording.

Digital audio involves a process whereby an audio waveform (such as the line output of a musical instrument) is sampled regularly and then converted into a series of binary words which actually represent the sound itself (see section 3.11.2). A digital audio recorder stores this sequence of data and can replay it by passing the original data through a digital-to-analogue convertor which turns the data back into a sound waveform, as shown in Figure 2.2. A multitrack recorder has a number of independent channels which work in the same way, allowing a sound recording to be built up in layers. MIDI, on the other hand, handles digital information which *controls* the generation of sound. MIDI data does not represent the sound waveform itself. When a multitrack music recording is made using a MIDI sequencer (see Chapter 5) this control data is stored, and can be replayed by transmitting the

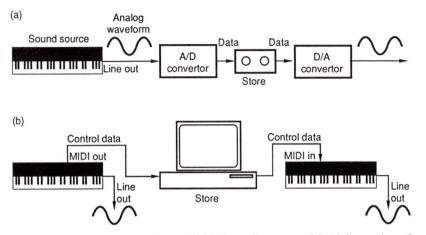

Figure 2.2 (a) Digital audio recording and (b) MIDI recording contrasted. In (a) the sound waveform itself is converted into digital data and stored, whereas in (b) only control information is stored, and a MIDI-controlled sound generator is required during replay

original data to a collection of MIDI-controlled musical instruments. It is the instruments which actually reproduce the recording.

A digital audio recording, then, allows any sound to be stored and replayed without the need for additional hardware. It is useful for recording acoustic sounds such as voices, where MIDI is not a great deal of help. A MIDI recording is almost useless without a collection of sound generators. An interesting advantage of the MIDI recording is that, since the stored data represents event information describing a piece of music, it is possible to change the music by changing the event data. MIDI recordings also consume a lot less memory space then digital audio recordings. It is also possible to transmit a MIDI recording to a different collection of instruments from those used during the original recording, thus resulting in a different sound. There is now a growing use of products which have integrated MIDI and digital audio recording within one software package, allowing the two to be edited and manipulated in parallel, and these are discussed in detail in section 5.9.

2.4 Basic MIDI principles

2.4.1 System specifications

The MIDI standard specifies a uni-directional serial interface running at 31.25 kbit/s ±1%. The rate was defined at a time when the clock speeds of microprocessors were typically much slower than they are today, this rate being a convenient division of the typical 1 or 2 MHz master clock rate. The rate had to be slow enough to be carried without excessive losses over simple cables and interface hardware, but fast enough to allow musical information to be transferred from one instrument to another without noticeable delays. Data is transmitted *uni-directionally,* that is from the transmitter to the receiver, and there is no return path unless a separate MIDI link is made.

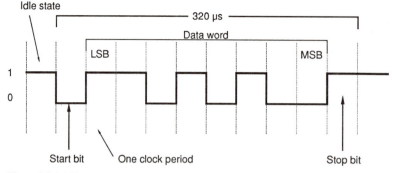

Figure 2.3 A MIDI message consists of a number of bytes, each transmitted serially and asynchronously by a UART in this format, with a start and stop bit to synchronise the receiving UART. The total period of a MIDI data byte, including start and stop bits, is 320 μs

Control messages are sent as groups of bytes, preceded by one start bit and followed by one stop bit per byte in order to synchronise reception of the data which is transmitted asynchronously, as shown in Figure 2.3. The addition of start and stop bits means that each 8 bit word actually takes ten bit periods to transmit (lasting a total of 320 μs) – a factor which must be borne in mind when attempting to calculate how long a particular message will take to transmit. Standard MIDI messages typically consist of one, two or three bytes, although there are longer messages for some purposes which will be covered later in this book.

2.4.2 The hardware interface

There is a defined hardware interface which should be incorporated in all MIDI equipment, which ensures electrical compatibility within a system. This is shown in Figure 2.4. Most equipment using MIDI has three interface connectors: *IN, OUT,* and *THRU* (sic). The OUT connector carries data which the device itself has generated. It originates from the device's UART (see section 1.8.5). The IN connector receives data from other devices and relays it to the device's UART. The THRU connector is a direct throughput of the data that is present at the IN. As can be seen from the hardware interface diagram, it is simply a buffered feed of the input data, and it has not been processed in any way. A few cheaper devices do not have THRU connectors, but it is possible to obtain 'MIDI THRU boxes' which provide a number of 'THRUs' from one input. Occasionally, devices without a THRU socket allow the OUT socket to be switched between OUT and THRU functions. A 5 mA current loop is created between a MIDI OUT or THRU and a MIDI IN, when connected with the appropriate cable, and data bits are signalled by the turning on and off of this current by the sending device. This principle is shown in Figure 2.5.

The interface incorporates an opto-isolator between the MIDI IN (that is the receiving socket) and the device's microprocessor system. This is to ensure that there is no direct electrical link between devices, and helps to reduce the effects of any problems which might occur if one instrument in a system were to develop an electrical fault. An opto-isolator is an encapsulated device in which a light-emitting diode (LED) can be turned on or off depending on the voltage applied across its terminals, illuminating a photo-transistor which consequently conducts or not,

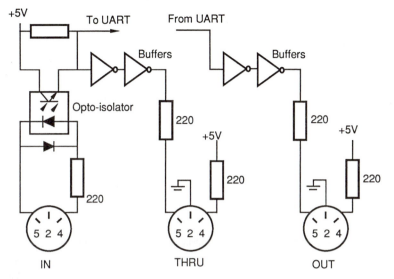

Figure 2.4 MIDI electrical interface showing IN, OUT and THRU ports

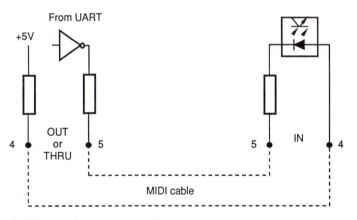

Figure 2.5 A current loop is formed between the OUT of the transmitter and the IN of the receiver when a MIDI cable is connected. The LED in the receiver's opto-isolator is turned on and off according to current flow

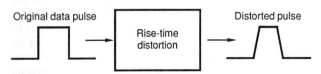

Figure 2.6 The edges of a square pulse subjected to rise-time distortion

depending on the state of the LED. Thus the data is transferred optically, rather than electrically. In the MIDI specification, the opto-isolator is defined as having a rise time of no more than 2 μs. The rise time affects the speed with which the device reacts to a change in its input, and if slow will tend to distort the leading edge of data bit cells, as shown in Figure 2.6. The same also applies in practice to fall times.

Rise-time distortion results in timing instability of the data, since it alters the time at which a data edge crosses the decision point between one and zero, and if the rise time is excessively slow the data value may be corrupted since the output of the device will not have risen to its full value before the next data bit arrives. If a large number of MIDI devices are wired in series (that is from THRU to IN a number of times) the data will be forced to pass through a number of opto-isolators and thus will suffer the combined effects of a number of stages of rise-time distortion. Whether or not this will be sufficient to result in data detection errors at the final receiver will depend to some extent on the quality of the opto-isolators concerned, and also on other losses which the signal may have suffered on its travels. It follows that the better the specification of the opto-isolator, the more stages of device cascading will be possible before unacceptable distortion is introduced.

Concerning any potential for delay between IN and THRU connectors, it should be stated at the outset that the delay in data passed through this circuit is only a matter of microseconds, so this contributes little to any audible delays perceived in the musical outputs of some instruments in a large system. The bulk of any perceived delay will be due to other factors, which are covered in later sections of this book.

2.4.3 Connectors and cables

The connectors used for MIDI interfaces are like the five-pin DIN plugs used in some hi-fi systems, and although it is possible to use hi-fi cables (depending on the way that they are wired), better quality connectors are to be preferred. The specification also allows for the use of XLR-type connectors (such as those used for balanced audio signals in professional equipment), although these are rarely encountered in practice. Only three of the pins of a 5 pin DIN plug are actually used in most equipment (the three innermost pins).

The cable should be a shielded twisted pair with the shield connected to pin 2 of the connector at both ends, although within the receiver itself, as can be seen from the diagram above, the MIDI IN does not have pin 2 connected to earth. This is to avoid earth loops, and makes it possible to use a cable either way round. (If two devices are connected together whose earths are at slightly different potentials, a current is caused to flow down any earth wire connecting them. This can induce interference into the data wires, possibly corrupting the data, and can also result in interference such as hum on audio circuits.)

It is recommended that no more than 15 m of cable is used for a single cable run in a simple MIDI system, and investigation of typical cables indicates that corruption of data does indeed ensue after longer distances, although this is gradual and depends on the electromagnetic interference conditions, the quality of cable and the equipment in use. Longer distances may be accommodated with the use of buffer or 'booster' boxes which act to compensate for some of the cable losses and retransmit the data. It is also possible to extend a MIDI system by using a networking approach, and this is discussed further in section 7.8.

In the cable, pin 5 at one end should be connected to pin 5 at the other, and likewise pin 4 to pin 4, and pin 2 to pin 2. Unless any hi-fi DIN cables to be used follow this convention they will not work. Professional microphone cable terminated in DIN connectors may be used as a higher-quality solution, because domestic cables will not always be a shielded twisted-pair and thus are more susceptible to external interference, as well as radiating more themselves which could interfere with adjacent audio signals. It is recommended that the correct cable is used in professional installations where MIDI cables are installed in the same trunking as audio cables, to avoid any problems with crosstalk.

2.4.4 Simple interconnection

Before going further it is necessary to look at a simple practical application of MIDI, to see how various aspects of the standard message protocol have arisen. For the present it will be assumed that we are dealing with a music system, although, as will be seen, MIDI systems can easily incorporate non-musical devices in various ways.

Having covered the *way* in which data is sent and received, it is now necessary to investigate *what* is sent and received, and how it relates to actions on the instruments themselves. Firstly, the functions of the standard interface sockets which exist on all MIDI devices. The OUT connector carries data representing actions which have taken place on the instrument itself: for example, it may carry data which says that certain keys were pressed or that a certain controller wheel was moved. The IN connector receives data from other equipment. Data messages received will be acted upon by the instrument concerned if it is capable of handling the particular message, and if it is destined for that instrument. The THRU connector, as described previously, is a direct throughput of the data that is present at the IN, and therefore does not include information about any of the actions which may have taken place on the instrument itself. The THRU socket can be used to 'daisy-chain' MIDI devices together, so that transmitted information from one controller can be sent to a number of receivers without the need for multiple outputs from a controller. Occasionally devices may have an internal 'merging' function which merges data from the device's front panel with data from the MIDI IN, sending the combined data to the MIDI OUT. (It is also possible to encounter devices which use the out connector for MIDI 'overflow', whereby any received note messages that exceed the device's polyphony are routed to the OUT for subsequent connection to daisy-chained modules.)

In the simplest MIDI system, one instrument could be connected to another as shown in Figure 2.7. Here, instrument 1 sends information relating to actions performed on its own controls (notes pressed, pedals pressed, etc.) to instrument 2, which imitates these actions as far as it is able. This type of arrangement can be used

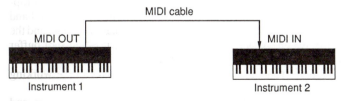

Figure 2.7 The simplest form of MIDI interconnection involves connecting two instruments together as shown

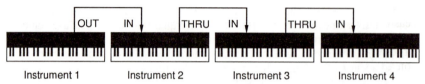

Figure 2.8 Further instruments can be added using THRU ports as shown, in order that messages from instrument 1 may be transmitted to all the other instruments

for 'doubling-up' sounds, 'layering' or 'stacking', such that a composite sound can be made up from two synthesisers' outputs. (It should be pointed out that the audio outputs of the two instruments would have to be mixed together for this effect to be heard.) Larger MIDI systems could be built up by further 'daisy-chaining' of instruments, such that instruments further down the chain all received information generated by the first (see Figure 2.8), although this is not a very satisfactory way of building a large MIDI system, and further methods of interconnection are discussed in section 7.6.

2.4.5 MIDI channels

Looking at the daisy-chained system in Figure 2.8, it will be appreciated that instruments 2, 3 and 4 all receive the information sent out by instrument 1. Instrument 1, in this case, could be considered as the controlling device, or master keyboard. The problem is how to define which instrument(s) in the chain will respond to which data messages.

It would be very useful if it were possible to specify on the transmitting instrument which receiving instrument was to respond, in order that a performer could control any of the other instruments simply by selecting a particular number on the master keyboard. If this were possible, then perhaps instruments 2, 3 and 4 need not even have keyboards of their own and could simply be 'black boxes' with MIDI interfaces, generating sound. In order to achieve this when there is only one piece of wire down which to send the information it is necessary to 'tag' each message with a label which says, in effect, 'this message is for instrument number three'. The other instruments would see the message, but they would also see the label indicating that the message was not for them. These tags are rather like addresses on letters: they ensure that the letters get to the right people.

MIDI messages are made up of a number of bytes. Each part of the message has a specific purpose, and one of these is to define the receiving channel to which the message refers. In this way, a controlling device can make data device-specific – in other words it can define which receiving instrument will act on the data sent. This is most important in large systems which use a computer sequencer as a master controller (see Chapter 5), when a large amount of information will be present on the MIDI data bus, not all of which is intended for every instrument. If a device is set in software to receive on a specific channel or on a number of channels it will act only on information which is 'tagged' with its own channel numbers. Everything else it will usually ignore. There are sixteen basic MIDI channels and instruments can usually be set to receive on any specific channel or channels (*omni off* mode), or to receive on all channels (*omni on* mode). The latter mode is useful as a means of determining whether anything at all is being received by the device.

Later it will be seen that the limit of sixteen MIDI channels can be exceeded easily by using multiport MIDI interfaces connected to a computer. In such cases it is important not to confuse the MIDI data channel with the physical port to which a device may be connected, since each physical port will be capable of transmitting on all sixteen data channels.

2.4.6 Message format

In order to see how the channel tag described above comes into play, it will be necessary to look at the MIDI message format in detail.

There are two basic types of MIDI message byte: the status byte and the data byte. Status bytes always begin with a binary one to distinguish them from data bytes, which always begin with a zero. As shown in Figure 2.9, the first half of the status byte denotes the message type and the second half denotes the channel number. Because the most significant bit (MSB) of each byte is reserved to denote the type (status or data) there are only seven active bits per byte which allows 2^7 (that is 128) possible values.

The first byte in a MIDI message is normally a status byte, which contains information about the channel number to which the message applies. It can be seen that four bits of the status byte are set aside to indicate the channel number, which allows for 2^4 (or 16) possible channels. The status byte is the label that denotes which receiver the message is intended for, and it also denotes which type of message is to follow (e.g.: a note on message). It will also be seen that there are three bits to denote the message type, because the first bit must always be a one. This theoreti-

Table 2.1 MIDI messages summarised

Message	Status	Data 1	Data 2
Note off	&8n	Note number	Velocity
Note on	&9n	Note number	Velocity
Polyphonic aftertouch	&An	Note number	Pressure
Control change (see Table 2.4)	&Bn	Controller number	Data
Program change	&Cn	Program number	–
Channel aftertouch	&Dn	Pressure	–
Pitch wheel	&En	LSbyte	MSbyte
System exclusive			
System exclusive start	&F0	Manufacturer ID	Data, (Data), (Data)
End of SysEx	&F7	–	
System common			
Quarter frame	&F1	Data (see Chapter 6)	–
Song pointer	&F2	LSbyte	MSbyte
Song select	&F3	Song number	–
Tune request	&F6	–	
System realtime			
Timing clock	&F8	–	–
Start	&FA	–	–
Continue	&FB	–	–
Stop	&FC	–	–
Active sensing	&FE	–	–
Reset	&FF	–	–

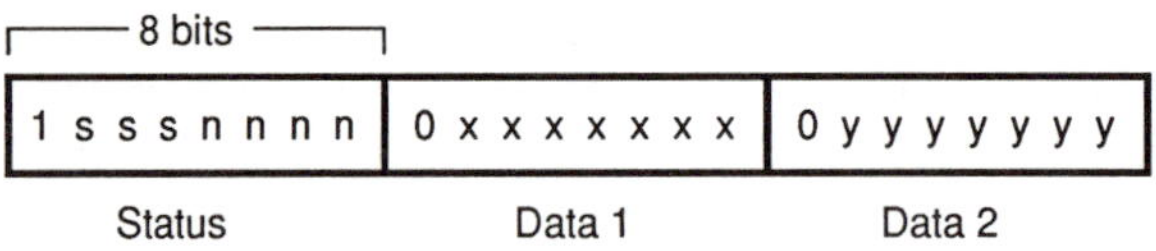

```
  ┌──── 8 bits ────┐
  ┌─────────────────┬─────────────────┬─────────────────┐
  │ 1 s s s n n n n │ 0 x x x x x x x │ 0 y y y y y y y │
  └─────────────────┴─────────────────┴─────────────────┘
       Status            Data 1            Data 2
```

Figure 2.9 General format of a MIDI message. The 'sss' bits are used to define the message type, the 'nnnn' bits define the channel number, whilst the 'xxxxxxx' and 'yyyyyyy' bits carry the message data. See text for details

cally allows for eight message types, but there are some special cases in the form of system messages (see below).

Standard MIDI messages can be up to three bytes long, but not all messages require three bytes, and there are some fairly common exceptions to the rule which are described below. Table 2.1 shows the format and content of MIDI messages under each of the statuses.

2.5 MIDI messages in detail

In this section the MIDI communication protocol will be examined in detail. The majority of the basic message types and their meanings will be explained, except for those dealing with timing and synchronisation which are covered in Chapter 6. The descriptions here are not intended as an alternative to reading the MIDI documentation itself, but rather as a commentary on it and an explanation of it. It follows that examples will be given, but that the reader should refer to the standard for a full listing of the protocol. This section will concentrate on issues relating to protocol and data format, whereas the chapters on implementation will go into more detail on the application of certain messages in real products. The prefix '&' will be used to indicate hexadecimal values throughout the discussion; individual MIDI message bytes will be delineated using square brackets, e.g. [&45], and channel numbers will be denoted using 'n' to indicate that the value may be anything from &0 to &F.

Recently the MMA and JMSC agreed to make a differentiation between the various addendums and modifications to the standard which have appeared since version 1.0 was formulated. These are known as Approved Protocols (APs) and Recommended Practices (RPs). An AP is a part of the standard MIDI specification, and is used when the standard is further defined or when a previously undefined command is defined, whereas an RP is used to describe an optional new MIDI application which is not a mandatory or binding part of the standard. Not all MIDI devices will have all the following commands implemented, since it is not mandatory for a device conforming to the MIDI standard to implement every possibility.

2.5.1 Channel and system messages contrasted

Two primary classes of message exist: those which relate to specific MIDI channels and those which relate to the system as a whole. One should bear in mind that it is possible for an instrument to be receiving in 'omni on' mode, in which case it would ignore the channel label and attempt to respond to anything that it received.

Channel messages start with status bytes in the range &8n to &En (they start at hexadecimal eight because the MSB must be a one for a status byte). System messages all begin with &F, and do not contain a channel number. Instead the least significant nibble of the system status byte (that which normally would contain the channel number) is used for further identification of the system message, such that there is room for sixteen possible system messages running from &F0 to &FF. System messages are themselves split into three groups: system common, system exclusive and system realtime. The common messages may apply to any device on the MIDI bus, depending only on the device's ability to handle the message. The exclusive messages apply to whichever manufacturer's devices are specified later in the message (see below), and the realtime messages are intended for devices which are to be synchronised to the prevailing musical tempo. (Some of the so-called realtime messages do not really seem to deserve this appellation, as discussed below.) The status byte &F1 is used for MIDI TimeCode (see Chapter 6).

2.5.2 Channel numbers

MIDI channel numbers are usually referred to as 'channels one to sixteen', but it can be appreciated that in fact the binary numbers which represent these run from zero to fifteen (&0 to &F), as fifteen is the largest decimal number which can be represented with four bits. Thus the note on message for channel 5 is actually &94 (nine for note on, and four for channel 5).

2.5.3 Note on and note off messages

Much of the musical information sent over a typical MIDI interface will consist of these two message types. As indicated by the titles, the note on message turns on a musical note, and the note off message turns it off. Note on takes the general format:

   [&8n] [Note number] [Velocity]

and note off takes the form:

   [&9n] [Note number] [Velocity] (although see section 2.5.4 below)

A MIDI instrument will generate note on messages at its MIDI OUT corresponding to whatever notes are pressed on the keyboard, on whatever channel the instrument is set to transmit. Also, any note which has been turned on must subsequently be turned off in order for it to stop sounding, thus if one instrument receives a note on message from another and then loses the MIDI connection for any reason, the note will continue sounding *ad infinitum*. This situation can occur if a MIDI cable is pulled out during transmission.

   MIDI note numbers relate directly to the western musical chromatic scale, and the format of the message allows for 128 note numbers which cover a range of a little over ten octaves – adequate for the full range of most musical material. This quantisation of the pitch scale is geared very much towards keyboard instruments, and is perhaps less suitable for other instruments and cultures where the definition of pitches is not so black-and-white. Nonetheless, means have been developed of adapting control to situations where unconventional tunings are required, as discussed in section 3.9. Note numbers normally relate to the musical scale as shown in Table 2.2, although there is a certain degree of confusion here. Yamaha established the use of C3 for middle C, whereas others have used C4. Some software allows the user to decide which convention will be used for display purposes.

Table 2.2 MIDI note numbers related to the musical scale

Musical note	MIDI note number
C–2	0
C–1	12
C0	24
C1	36
C2	48
C3 (middle C)	60 (Yamaha convention)
C4	72
C5	84
C6	96
C7	108
C8	120
G8	127

2.5.4 Velocity information

Note messages are associated with a velocity byte, and this is used to represent the speed at which a key was pressed or released. The former will correspond to the force exerted on the key as it is depressed: in other words, 'how hard you hit it' (called 'note on velocity'). It is used to control parameters such as the volume or timbre of the note at the audio output of an instrument, and can be applied internally to scale the effect of one or more of the envelope generators in a synthesiser. This velocity value has 128 possible states, but not all MIDI instruments are able to generate or interpret the velocity byte, in which case they will set it to a value half way between the limits, i.e.: 64_{10}. Some instruments may act on velocity information even if they are unable to generate it themselves. It is recommended that a logarithmic rather than linear relationship should be established between the velocity value and the parameter which it controls, since this corresponds more closely to the way in which musicians expect an instrument to respond, although some instruments allow customised mapping of velocity values to parameters.

Note off velocity (or 'release velocity') is not widely used, as it relates to the speed at which a note is released, which is not a parameter that affects the sound of many normal keyboard instruments. Nonetheless it is available for special effects if a manufacturer decides to implement it.

The note on, velocity zero value is reserved for the special purpose of turning a note off, for reasons which will become clear under 'Running status' below. If an instrument sees a note number with a velocity of zero, its software should interpret this as a note off message.

2.5.5 Running status

When a large amount of information is transmitted over a single MIDI bus, delays naturally arise due to the serial nature of transmission wherein data such as the concurrent notes of a chord must be sent one after the other. It will be advantageous, therefore, to reduce the amount of data transmitted as much as possible, in order to keep the delay as short as possible and to avoid overloading the devices on the bus with unnecessary data.

Running status is an accepted method of reducing the amount of data transmitted, and one which all MIDI software should understand. It involves the assumption that once a status byte has been asserted by a controller there is no need to reiterate this status for each subsequent message of that status, so long as the status has not changed in between. Thus a string of note on messages could be sent with the note on status only sent at the start of the series of note data, for example:

[&9n] [Data] [Velocity] [Data] [Velocity] [Data] [Velocity]

It will be appreciated that for a long string of note data this could reduce the amount of data sent by nearly one third. But as in most music each note on is almost always followed quickly by a note off for the same note number, this method would clearly break down, as the status would be changing from note on to note off very regularly, thus eliminating most of the advantage gained by running status. This is the reason for the adoption of note on, velocity zero as equivalent to a note off message, because it avoids a change of status during running status, allowing a string of what appears to be note on messages, but which is, in fact, both note on and note off.

Running status is not used at all times for a string of same-status messages, and will often only be called upon by an instrument's software when the rate of data exceeds a certain point. Indeed, an examination of the data from a typical synthesiser indicates that running status is not used during a large amount of ordinary playing. Yet it might be useful for a computer sequencer which can record data from a large number of MIDI devices, and might transmit it all out of a single port on replay. Even so, the benefit would not be particularly great even in this case, as the sequencer would be alternating between the addressing of a number of different devices with different statuses, in order not to allow any one device to lag behind in relation to the others.

2.5.6 Polyphonic key pressure (aftertouch)

The key pressure messages are sometimes called 'aftertouch' by keyboard manufacturers. Aftertouch is perhaps a slightly misleading term as it does not make clear what aspect of touch is referred to, and many people have confused it with note off velocity. This message refers to the amount of pressure placed on a key at the bottom of its travel, and it is used to instigate effects based on how much the player leans onto the key after depressing it. It is often applied to performance parameters such as vibrato.

The polyphonic key pressure message is not widely used, as it transmits a separate value for every key on the keyboard, and thus requires a separate sensor for every key. This can be expensive to implement, and is beyond the scope of many keyboards, so most manufacturers have resorted to the use of the channel pressure message (see below). It should be noted, though, that some manufacturers have shown it to be possible to implement this feature at a reasonable cost. The message takes the general format:

[&An] [Note number] [Pressure]

Implementing polyphonic key pressure messages involves the transmission of a considerable amount of data over MIDI which might well be unnecessary, as the message will be sent for every note in a chord every time the pressure changes. As most people do not maintain a constant pressure on the bottom of a key whilst playing, many messages might be sent per note. A technique known as 'controller thinning' may be used by a device to limit the rate at which such messages are transmitted, and this may be implemented either before transmission or at a later stage using a computer. Alternatively this data may be filtered out altogether if it is not required.

2.5.7 Control change

As well as note information, a MIDI device may be capable of transmitting control information which corresponds to the various switches, control wheels and pedals associated with it. These come under the control change message group, and should be distinguished from program change messages (see section 2.5.9). The controller messages have proliferated enormously since the early days of MIDI, and not all devices will implement all of them. The control change message takes the general form:

[&Bn] [Controller number] [Data]

thus a number of controllers may be addressed using the same type of status byte by changing the controller number.

Although the original MIDI standard did not lay down any hard and fast rules for the assignment of physical control devices to logical controller numbers, there is now common agreement amongst manufacturers that certain controller numbers will be used for certain purposes, and these are controlled by the MMA and JMSC. It should be noted that there are two distinct kinds of controller: that is, the switch type, and the analogue type. The analogue controller is any continuously variable wheel, lever, slider or pedal that might have any one of a number of positions, and these are often known as continuous controllers. There are 128 controller numbers available, and these are grouped as shown in Table 2.3. Table 2.4 shows a more detailed breakdown of the use of these, as found in the majority of MIDI-controlled musical instruments, although the full list is regularly updated by the MMA. The control change messages have become fairly complex so coverage of them is divided into a number of sections. The topics of sound control, bank select and effects control have been left for coverage by later chapters on MIDI implementation in synthesisers and effects devices.

The first 64 controller numbers (that is up to &3F) relate to only 32 *physical* controllers (the continuous controllers). This is to allow for greater resolution in the quantisation of position than would be feasible with the seven bits that are offered by a single data byte. Seven bits would only allow 128 possible positions of an analogue controller to be represented, and it was considered that this might not be adequate in some cases. For this reason, the first 32 controllers handle the most significant byte (MSbyte) of the controller data, while the second 32 handle the least significant byte (LSbyte). In this way, controller numbers &06 and &38 both represent the data entry slider, for example. Together, the data values can make up a 14 bit number (because the first bit of each data word has to be a zero), which allows the quantisation of a control's position to be one part in 2^{14} (16384_{10}). Clearly, not all controllers will require this resolution, but it is available if needed. Only the LSbyte would be needed for small movements of a control. If a system opts not to use the extra resolution offered by the second byte, it should send only the MSbyte for coarse control, and in practice this is all that is transmitted on many devices.

On/off switches can be represented easily in binary form (0 for OFF, 1 for ON), and it would be possible to use just a single bit for this purpose, but, in order to conform to the standard format of the message, switch states are normally repre-

Table 2.3 MIDI controller classifications

Controller number (hex)	Function
&00–1F	14 bit controllers, MSbyte
&20–3F	14 bit controllers, LSbyte
&40–65	7 bit controllers or switches
&66–77	Currently undefined
&78–7F	Channel mode control

Table 2.4 MIDI controller functions

Controller number (hex)	Function
00	Bank select
01	Modulation wheel
02	Breath controller
03	Undefined
04	Foot controller
05	Portamento time
06	Data entry slider
07	Main volume
08	Balance
09	Undefined
0A	Pan
0B	Expression controller
0C	Effect control 1
0D	Effect control 2
0E–0F	Undefined
10–13	General purpose controllers 1–4
14–1F	Undefined
20–3F	LSbyte for 14 bit controllers (same function order as 00–1F)
40	Sustain pedal
41	Portamento on/off
42	Sostenuto pedal
43	Soft pedal
44	Legato footswitch
45	Hold 2
46–4F	Sound controllers (see Chapter 3)
50–53	General purpose controllers 5–8
54	Portamento control
55–5A	Undefined
5B–5F	Effects depth 1–5 (see Chapter 3)
60	Data increment
61	Data decrement
62	NRPC LSbyte
63	NRPC MSbyte
64	RPC LSbyte
65	RPC MSbyte
66–77	Undefined
78	All sounds off
79	Reset all controllers
7A	Local on/off
7B	All notes off
7C	Omni receive mode off
7D	Omni receive mode on
7E	Mono receive mode
7F	Poly receive mode

sented by data values between &00 and &3F for OFF, and &40–&7F for ON. In other words switches are now considered as 7 bit continuous controllers, and it may be possible on some instruments to define positions in between off and on in order to provide further degrees of control, such as used in some 'sustain' pedals (although this is not common in the majority of equipment). In older systems it may be found that only &00 = OFF and &7F = ON.

The data increment and decrement buttons which are present on many devices are assigned to two specific controller numbers (&60 and &61), and a more recent extension to the standard defines four controllers (&62 to &65) which effectively expand the scope of the control change messages. These are the registered and non-registered parameter controllers (RPCs and NRPCs), discussed in section 3.7.3.

The 'all notes off' command (frequently abbreviated to 'ANO') was designed to be transmitted to devices as a means of silencing them, but it does not necessarily have this effect in practice. What actually happens varies between instruments, especially if the sustain pedal is held down or notes are still being pressed manually by a player. All notes off is supposed to put all note generators into the release phase of their envelopes, and clearly the result of this will depend on what a sound is programmed to do at this point. The exception should be notes which are being played while the sustain pedal is held down, which should only be released when that pedal is released. 'All sounds off' was designed to overcome the problems with 'all notes off', by turning sounds off as quickly as possible.

'Reset all controllers' is designed to reset all controllers to their default state, in order to return a device to its 'standard' setting.

2.5.8 Channel modes

Although grouped with the controllers, under the same status, the channel mode messages differ somewhat in that they set the mode of operation of the instrument receiving on that particular channel.

'Local on/off' is used to make or break the link between an instrument's keyboard and its own sound generators. Effectively there is a switch between the

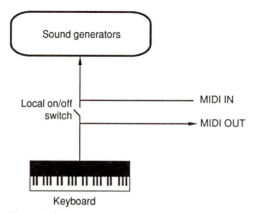

Figure 2.10 The 'local off' switch disconnects a keyboard from its associated sound generators in order that the two parts may be treated independently in a MIDI system

output of the keyboard and the control input to the sound generators which allows the instrument to play its own sound generators in normal operation when the switch is closed (see Figure 2.10). If the switch is opened, the link is broken and the output from the keyboard feeds the MIDI OUT while the sound generators are controlled from the MIDI IN. In this mode the instrument acts as two separate devices: a keyboard without any sound, and a sound generator without a keyboard. This configuration can be useful when the instrument in use is the master keyboard for a large sequencer system, where it may not always be desired that everything played on the master keyboard results in sound from the instrument itself.

'Omni off' ensures that the instrument will only act on data tagged with its own channel number(s), as set by the instrument's controls. 'Omni on' sets the instrument to receive on all of the MIDI channels. In other words, the instrument will ignore the channel number in the status byte and will attempt to act on any data that may arrive, whatever its channel. Devices should power-up in this mode according to the original specification, but more recent devices will tend to power up in the mode that they were left. Mono mode sets the instrument such that it will only reproduce one note at a time, as opposed to 'Poly' (phonic) in which a number of notes may be sounded together.

In older devices the mono mode came into its own as a means of operating an instrument in a 'multitimbral' fashion, whereby MIDI information on each channel controlled a separate monophonic musical voice. This used to be one of the only ways of getting a device to generate more than one type of voice at a time. The data byte that accompanies the mono mode message specifies how many voices are to be assigned to adjacent MIDI channels, starting with the basic receive channel. For example, if the data byte is set to 4, then four voices will be assigned to adjacent MIDI channels, starting from the basic channel which is the one on which the instrument has been set to receive in normal operation. Exceptionally, if the data byte is set to zero, all sixteen voices (if they exist) are assigned each to one of the sixteen MIDI channels. In this way, a single multitimbral instrument can act as sixteen monophonic instruments, although on cheaper systems all of these voices may be combined to one audio output.

Mono mode tends to be used mostly on MIDI guitar synthesisers since each string can then have its own channel, and each can control its own set of pitch bend and other parameters. The mode also has the advantage that it is possible to play in a truly legato fashion – that is with a smooth take over between the notes of a melody – because the arrival of a second note message acts simply to change the pitch if the first one is still being held down, rather than re-triggering the start of a note envelope. The legato switch controller (see Table 2.4) allows a similar type of playing in polyphonic modes by allowing new note messages only to change the pitch

In poly mode the instrument will sound as many notes as it is able at the same time. Instruments differ as to the action taken when the number of simultaneous notes is exceeded: some will release the first note played in favour of the new note, whereas others will refuse to play the new note. Some may be able to route excess note messages to their MIDI OUT ports so that they can be played by a chained device. The more intelligent of them may look to see if the same note already exists in the notes currently sounding, and only accept a new note if is not already sounding. Even more intelligently, some devices may release the quietest note (that with the lowest velocity value), or the note furthest through its velocity envelope, to make way for a later arrival. It is also common to run a device in poly mode on more than one receive channel, provided that the software can handle the reception of multiple polyphonic channels. A multitimbral sound generator may well have this facility, commonly referred to as 'multi' mode, making it act as if it were a number of separate instruments each receiving on a separate channel. In multi mode a device may be able to dynamically assign its polyphony between the channels and voices in order that the user does not need to assign a fixed polyphony to each voice.

It should be noted that a change of channel mode should turn all notes off automatically.

2.5.9 Program change

The program change message is used most commonly to change the 'patch' of an instrument or other device. A patch is a stored configuration of the device, describing the setup of the tone generators in a synthesiser, and the way in which they are interconnected. Program change is channel-specific, and there is only a single data byte associated with it, specifying to which of 128 possible stored programs the receiving device should switch. On non-musical devices such as effects units, the program change message is often used to switch between different effects, and the different effects programs may be mapped to specific program change numbers. The message takes the general form:

&[Cn] [Program number]

If a program change message is sent to a musical device, it will usually result in a change of voice, as long as this facility is enabled. Exactly which voice corresponds to which program change number depends on the manufacturer, and this is usually specified in the manual. It is quite common for some manufacturers to implement this function in such a way that a data value of zero gives voice number one. This results in a permanent offset between the program change number and the voice number, which should be taken into account in any software. On some instruments, voices may be split into a number of 'banks' of 8, 16 or 32, and higher banks can be selected over MIDI by setting the program change number to a value which is 8, 16 or 32 higher than the lowest bank number. For example, bank 1, voice 2, might be selected by program change &01, whereas bank 2, voice 2, would probably be selected in this case by program change &11, where there were sixteen voices per bank.

There are also a number of other approaches used in commercial sound modules, described in section 3.8. Where more than 128 voices need to be addressed remotely, the more recent 'bank select' command may be implemented.

2.5.10 Channel aftertouch

As explained under 'Polyphonic key pressure' it is rarely economical to send a message representing the individual pressure applied at the bottom of every key pressed. Thus most instruments use a single sensor, often in the form of a pressure-sensitive conductive plastic bar running the length of the keyboard, to detect the pressure applied to keys at the bottom of their travel. In the case of channel aftertouch, one message is sent for the entire instrument, and this will correspond to an approximate total of the pressure over the range of the keyboard, the strongest influence being from the key pressed the hardest. (Some manufacturers have split the pressure detector into upper and lower keyboard regions, and some use 'intelligent' zoning.) The message takes the general form:

&[Dn] [Pressure value]

There is only one data byte, thus there are 128 possible values, and, as with the polyphonic version, many messages may be sent as the pressure is varied at the bottom of a key's travel. Controller 'thinning' may be used to reduce the quantity of these messages, as described above.

2.5.11 Pitch bend wheel

The pitch wheel message has a status byte of its own, and carries information about the movement of the sprung-return control wheel on many keyboards which modifies the pitch of any note(s) played. It uses two data bytes in order to give 14 bits of resolution, in much the same way as the continuous controllers, except that the pitch wheel message carries both bytes together. Fourteen data bits are required so that the pitch appears to change smoothly, rather than in steps (as it might with only seven bits). It should be noted that the pitch bend message is channel specific, and thus ought to be sent separately for each individual channel. This becomes important when using a single multi-timbral device in mono mode (see above), as one must ensure that a pitch bend message only affects the notes on the intended channel. The message takes the general form:

&[En] [LSbyte] [MSbyte]

The value of the pitch bend controller should be halfway between the lower and upper range limits when it is at rest in its sprung central position, thus allowing bending both down and up. This corresponds to a hex value of &2000, transmitted as &[En] [00] [40]. The range of pitch controlled by the bend message is set on the receiving device itself, or using the RPC designated for this purpose (see section 3.7.3)

2.5.12 System exclusive

A system exclusive message is one which is unique to a particular manufacturer, and often a particular instrument. The only thing that is defined about such messages is how they are to start and finish, with the exception of the use of system exclusive messages for universal information, as discussed elsewhere. System exclusive messages generated by a device will naturally be produced at the MIDI OUT, not at the THRU, so a deliberate connection must be made between the transmitting device and the receiving device before data transfer may take place. Occasionally it is necessary to make a return link from the OUT of the receiver to the IN of the transmitter so that two-way communication is possible, and so that the receiver can control the flow of data to some extent by telling the transmitter when it is ready to receive and when it has received correctly (a form of handshaking).

The message takes the general form:

&[F0] [ident.] [data] [data] ... [F7]

where [ident.] identifies the relevant manufacturer ID, which is a number defining which manufacturer's message is to follow. Originally, manufacturer IDs were a single byte but the number of IDs has been extended by setting aside the [00] value of the ID to indicate that two further bytes of ID follow. Manufacturer IDs are therefore either one or three bytes long. A full list of manufacturer IDs is available from the MMA.

Data of virtually any sort can follow the ID. It can be used for a variety of miscellaneous purposes which have not been defined in the MIDI standard, and the message can have virtually any length that the manufacturer requires, although it is often split into packets of a manageable size in order not to cause receiver memory buffers to overflow. Exceptions are data bytes which look like other MIDI status bytes (except realtime messages), as they will naturally be interpreted as such by any receiver, which might terminate reception of the system exclusive message. The message should be terminated with &F7, although this is not always observed, in which case the receiving device should 'time-out' after a given period, or terminate the system exclusive message on receipt of the next status byte. It is recommended that some form of error checking (typically a checksum) is employed for long system exclusive data dumps, and many systems employ means of detecting whether the data has been received accurately, asking for re-tries of sections of the message in the event of failure, via a return link to the transmitter.

Examples of applications for such messages can be seen in the form of sample data dumps (from a sampler to a computer and back again for editing purposes), although this is painfully slow, and voice data dumps (from a synthesiser to a computer for storage and editing of user-programmed voices). There are now an enormous number of uses of system exclusive messages, both in the universal categories and in the manufacturer categories, and these will be introduced as and when appropriate in the course of this book. It is not intended to list every possible implementation, and the reader is referred to other publications which have made this a speciality (see Appendix 2).

2.5.13 Universal system exclusive messages

The three highest numbered IDs within the system exclusive message have been set aside to denote special modes. These are the 'universal non-commercial' messages (ID: &7D), the 'universal non-realtime' messages (ID: &7E), and the 'universal realtime' messages (ID: &7F). Specific uses of the universal messages will be introduced in appropriate sections later in the book.

Universal non-commercial messages are set aside for educational and research purposes and should not be used in commercial products. Universal non-realtime messages are used for universal system exclusive events which are not time critical, and universal realtime messages deal with time critical events (thus being given a higher priority). The two latter types of message normally take the general form of:

&[F0] [ID] [dev. ID] [sub-ID #1] [sub-ID #2] [data] ... ... [F7]

Device ID used to be referred to as 'channel number', but this did not really make sense since a whole byte allows for the addressing of 128 channels, and this does not correspond to the normal 16 channels of MIDI. The term 'device ID' is now used widely by software as a means of defining one of a number of physical devices in a large MIDI system, rather than defining a MIDI channel number. It should be noted, though, that it is allowable for a device to have more than one device ID if this seems appropriate. Modern MIDI devices will normally allow their device ID to be set either over MIDI or from the front panel. The use of &7F in this position signifies that the message applies to all devices as opposed to just one.

The sub-IDs are used to identify firstly the category or application of the message (sub-ID #1) and secondly the type of message within that category (sub-ID #2). For some reason, the original MIDI sample dump messages (see section 3.14.2) do not use the sub-ID #2, although some recent additions to the sample dump do.

2.5.14 Tune request

Older analogue synthesisers tended to drift somewhat in pitch over the time that they were turned on. The tune request is a request for these synthesisers to re-tune themselves to a fixed reference. (It is advisable not to transmit pitch bend or note on messages to instruments during a tune up because of the unpredictable behaviour of some products under these conditions.)

2.5.15 Active sensing

Active sensing messages are single status bytes sent roughly three times per second by a controlling device when there is no other activity on the bus. It acts as a means of reassuring the receiving devices that the controller has not disappeared. Not all devices transmit active sensing information, and a receiver's software should be able to detect the presence or lack of it. If a receiver has come to expect active sensing bytes then it will generally act by turning off all notes if these bytes disappear for any reason. This can be a useful function when a MIDI cable has been pulled out during a transmission, as it ensures that notes will not be left sounding for very long. If a receiver has not seen active sensing bytes since last turned on, it should assume that they are not being used.

2.5.16 Reset

This message resets all devices on the bus to their power-on state. The process may take some time and some devices mute their audio outputs, which can result in clicks, therefore the message should be used with care.

2.6 Message priority

A standard priority should exist in the handling of message types which ensures that the most urgent are handled first. The order is:
Reset
System exclusive
System realtime
System common and channel messages
Within the system exclusive category, universal realtime messages take the highest priority because they are required for synchronisation purposes and may be time critical. Furthermore, system realtime messages may be allowed to interrupt a normal MIDI message (even a system exclusive message) in between any pair of bytes, in order for the timing to be kept as stable as possible. For example, a timing clock byte [&F8], as discussed in Chapter 6, may be inserted by a computer between the two data bytes of a note on message if necessary, without disturbing the current note on status. It is not necessary to reiterate the original status after the realtime byte has passed, and receiving software should revert automatically.

Chapter 3

Implementing MIDI in musical instruments

This chapter will examine the way in which MIDI relates to the control of musical sound generators such as synthesisers and samplers. It is not intended to provide in-depth coverage of the way in which synthesisers and samplers work, although an overview will be given.

3.1 An introduction to synthesisers

3.1.1 Background

A synthesiser is an electronic musical instrument capable of creating sounds artificially. In its earliest days a synthesiser might have occupied a large room, or at least a few racks, whereas today it may be housed in a tiny rack-mounted box having only a few buttons and a miniscule display. There has also been a significant change in the use of synthesisers, since the early devices were really experimental systems based on analogue technology such as oscillators, filters and amplifiers, in which sounds were created by physically connecting one section so as to modulate another, for example, or routing a particular oscillator through a particular filter. Little or no facility existed for storing patch configurations except that of writing them down. These synthesisers would probably have been limited in the number of simultaneous sounds they could produce, often even being monophonic (one note at a time). More recent synthesisers are based on digital sound processing, have a built-in computer, and incorporate facilities for memorising large banks of internal configurations or 'patches'. They may come from the factory preset with a number of sounds that have already been created, to the extent that many people never get involved at all with the actual process of sound synthesis. They are polyphonic and multi-timbral (see section 3.1.3) and can be remotely controlled using MIDI.

One now talks in terms of 'sound modules' rather than synthesisers, and such sound modules often incorporate a combination of 'real' sounds (sampled sounds) and electronically synthesised sounds. Commonly the sound module has no keyboard because it is only going to be controlled using a MIDI interface, and many have few front panel controls for the same reason. The most recent sound modules typically have stereo outputs and often incorporate built-in effects units for adding reverberation to internal voices.

3.1.2 Voice generation

Although the process by which synthesiser voices are generated varies, the block diagram in Figure 3.1 shows a generalised process. A source generates a sound waveform which is subsequently modified in either its spectral content, its amplitude or both. Optionally certain effects may be added afterwards. Each stage of the process may be controlled externally to a varying extent. A note's pitch is controlled by affecting the source waveform's frequency, its timbre is controlled by affecting the spectral content, and its volume is affected by changing the amplitude. Any or all of these processes can be made to change in a controlled manner over a period of time, and triggered remotely.

In a modern sound module the wave source is usually a digital wave table stored in memory, whose numerical values represent sample amplitudes of a basic cycle of a wave, or which may be a cycle or two of a sampled sound, or even the complete sampled sound. (Basic digital sampling is described in section 3.11.) When a note is played the wave samples are read out from the memory at a rate corresponding to the pitch of the note, and the pitch of the note is altered by varying the sampling rate. Alternatively, a process of pitch shifting involving interpolation (adding calculated sample values at points between the original samples) and decimation (removing samples in a similar way) may be used. The wave table is read over and over again as the note is held down. This process fulfils much the same function as an oscillator, and each 'oscillator' is capable of producing a single note at a time.

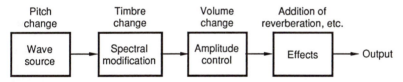

Figure 3.1 Generalised block diagram of sound generation in a synthesiser

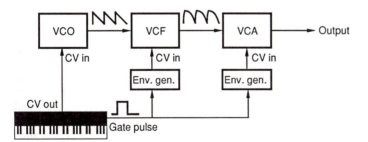

Figure 3.2 In this section of an analogue synthesiser a tone of variable pitch is generated by a VCO (voltage controlled oscillator), filtered by a VCF (voltage controlled filter) and passed through a VCA (voltage controlled amplifier) to alter its amplitude. Envelope generators, producing a DC output which varies with time, influence the characteristics of the voltage controlled sections in order to change the sound. The start of the envelope is triggered by a pulse from the keyboard whose leading edge normally corresponds to the 'note on' point. Other control signals may be applied to the CV inputs of each stage, such as joystick controls or a low frequency oscillator

At least as many oscillators are required as the number of notes to be produced simultaneously, and often many more are used in order either to modulate or add to each other to produce complex waves. In FM synthesis for example, as implemented by Yamaha, oscillators are called 'operators' and may be configured in various 'algorithms' so that an operator either frequency-modulates the next one in the chain or is added to the output of another. The result can be a wave with very complex characteristics due to the compounded stages of modulation.

Each stage of the synthesis chain may be externally influenced, as introduced above. In earlier analogue synthesisers this involved using DC voltage control to change the frequency of oscillators, the characteristics of filters and the gain of amplifiers, as shown in Figure 3.2. An external keyboard would produce a voltage proportional to the note pressed and this would be connected to the voltage-controlled oscillator (VCO) in order to affect the pitch of the sound. It was common to work to a standard of 1 volt per octave of pitch. In order to affect the way in which sound characteristics changed with time, envelope generators were used, connected to the various voltage-controlled elements of the synthesiser and producing a varying DC voltage according to a prearranged algorithm. The concept is still very much the same today, except that instead of DC voltage control the process is one of digital control, whereby binary values are used as multipliers for signal processing functions which generate similar effects.

Envelope generators produce a time-varying output in a number of stages, as illustrated in Figure 3.3. This typical four-stage envelope has attack, decay, sustain and release phases, and could be used to control any internal function of a synthesiser such as pitch, timbre or volume. Each phase can be programmed to have a certain maximum level and a certain rate (that is the speed with which it reaches the maximum level). The envelope cycle is normally started when a note is pressed, upon which it enters the attack phase, after which it decays to the sustain portion. This sustain portion is held while the note is held down and only passes to the release phase when the note is lifted. The audible effect resulting from changes in envelope shape depend on the parameter to which the envelope is linked.

In FM synthesis for example, as shown in Figure 3.4, each operator has its own independent envelope generator which scales the way in which its amplitude

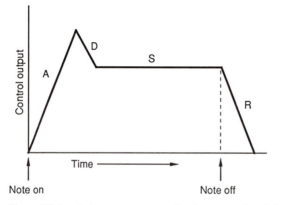

Figure 3.3 A typical envelope generator has four stages. Attack (A), Decay (D), Sustain (S) and Release (R). The rates and maximum values of each stage of the envelope can be set independently

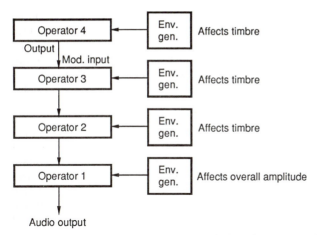

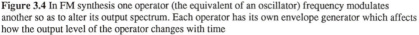

Figure 3.4 In FM synthesis one operator (the equivalent of an oscillator) frequency modulates another so as to alter its output spectrum. Each operator has its own envelope generator which affects how the output level of the operator changes with time

changes with time. Those higher up the modulation chain will affect the timbre, whereas the one at the bottom will affect the overall amplitude of the output of that part of the algorithm. Commonly, envelope generators can be made velocity sensitive in order that their effect may be limited by the velocity value of a note. In this way the characteristics of a note, such as 'brightness', can be affected during performance.

MIDI control provides a means by which many of these parameters may be externally triggered or accessed. A note on message, for example, contains a number of pieces of information. Firstly it indicates which channel is concerned (which may be used to select the appropriate voice to be played), then it indicates the note number which is used to control the pitch of the sound, and then it indicates the velocity with which the note was depressed (which could be applied to an envelope generator to influence one of a variety of sound parameters). The note on message also triggers the envelope generators connected with the note to be played so that they are reset to the starts of their cycles. The way in which MIDI commands are mapped to voice generation parameters is part of the internal setup of the sound module or synthesiser, and may be changed to some extent by the user. Various example are given in the remainder of this chapter.

3.1.3 Polyphony, voice and note assignment

Although many older synthesisers were monophonic (in other words, they could only generate one note at a time), it is now more common for a synthesiser to be polyphonic (capable of generating many notes together). Nonetheless, a polyphonic device is usually capable of operating as a monophonic synthesiser if set to that mode either over MIDI or from the front panel. It may be easier to obtain true legato melodies and portamento effects by playing monophonically, as the synthesiser will not allow two notes to sound simultaneously, and will normally sound the most recent note even if another is still depressed.

Modern sound modules tend to be at least 16 note polyphonic, and many allow considerably greater numbers of simultaneous notes to be sounded. This becomes particularly important when considering the dynamic note assignment that takes place in multi-timbral sound modules (see below). When the polyphony of a device is exceeded the device should follow a predefined set of rules to determine what to do with the extra notes. Typically a sound module will either release the 'oldest' notes first, or possibly release the quietest. Alternatively, new notes which exceed the polyphony will simply not be sounded until others are released.

It is important to distinguish between the degree of polyphony offered by a device and the number of simultaneous *voices* it can generate. Sometimes these may be traded off against each other in multi-timbral devices (see below), by allocating a certain number of notes to each voice, with the total adding up to the total polyphony. You could then have, say, either all 16 notes allocated to one voice or 4 notes to each of 4 voices. Dynamic allocation is often used to distribute the polyphony around the voices depending on demand, and this is a particular feature of General MIDI sound modules (see section 3.8.3).

A multi-timbral sound generator is one which is capable of generating more than one voice at a time, independent of polyphony considerations. A voice is a particular sound type, such as 'grand piano' or 'accordion'. This capability is now the norm for modern sound modules. Older synthesisers used to be able to generate only one or two voices at a time, possibly allowing a keyboard split, and could sometimes make use of MIDI channel mode 4 (see below) to allow multiple *monophonic* voices to be generated under MIDI control. They tended only to receive polyphonically on one MIDI channel at a time. More recent systems are capable of receiving on all 16 MIDI channels simultaneously, with each channel controlling an entirely independent polyphonic voice.

3.1.4 Onboard sequencing

Some synthesisers have their own built-in sequencers, making them into stand-alone music production systems. MIDI is not strictly necessary in such cases, but often one may need to integrate such synthesisers with a larger MIDI system, perhaps controlled by an overall sequencer. In such cases it is necessary to decide whether the onboard sequencer is actually necessary, or whether in fact it would be more straightforward to control and record everything on a separate computer. Alternatively, the onboard sequencer could be used to add to the total capabilities of the system, and would need to be able to lock either to MTC, or to MIDI clock and song pointer data in order to be able to run in sync with the master sequencer (see Chapter 6). Onboard sequencers can be used as a form of 'notepad' for trying out ideas, after which the data may be transferred to an external sequencer for editing.

If the onboard sequencer is to be used to control sound generators other than those within the synth, then it must be possible actually to transmit MIDI control data from the synth's MIDI out for this purpose. Some internal sequencers only control the synth itself and do not operate over MIDI.

3.1.5 Audio inputs and outputs

A synth or sound module may have more than one audio output. The most basic systems have a single mono output (not to be confused with monophonic sound

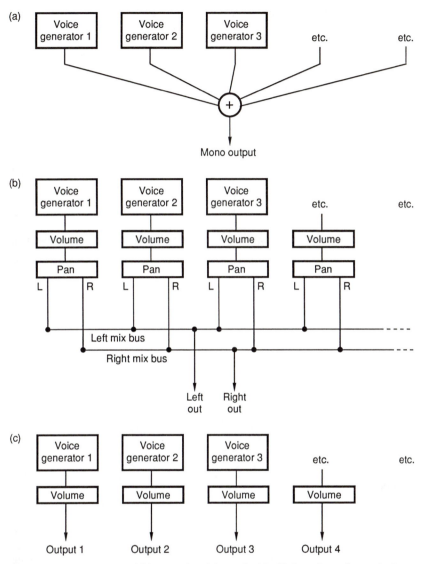

Figure 3.5 Voice generators within a sound module may be (a) added together to form a single mono output, (b) internally mixed and panned to form a single stereo output, or (c) fed to separate outputs. It is also possible to combine these approaches so that a device with a stereo output also has individual outputs, for example

generation), and all sound sources within the device are mixed together as shown in Figure 3.5. More common these days is for devices to have stereo outputs, and in such devices it is possible for internal voices to be panned to a point between left and right. In simpler systems a voice may have to be assigned to either left, right or centre (routed to both left and right). It may also be possible to control the overall volume of each voice, which together with internal pan control really results in a

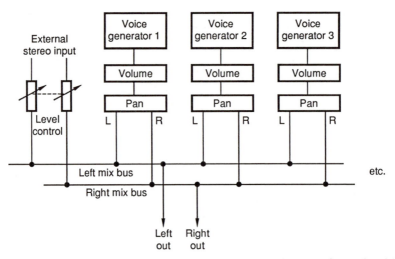

Figure 3.6 An external stereo input may be internally mixed with the output of a sound module

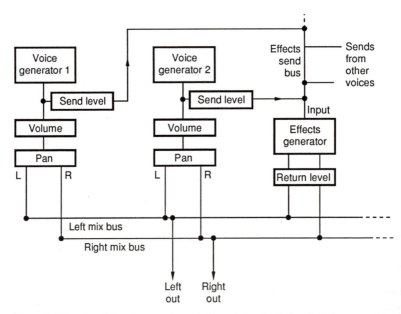

Figure 3.7 Sound module voices may be routed to an internal effects unit. Only one send bus is shown here, therefore only one effect is possible at a time. The send level controls alter the amount of each voice being fed to the effects unit (the send is normally monophonic). The return level control adjusts the level of the effects in the stereo mix output

miniature stereo mixer within the sound module, often eliminating the need for external mixing.

More expensive sound generators may have multiple audio outputs as well as a stereo output. Using these it is possible for each voice to have its own output, and this allows individual voices to be routed to individual channels on an external audio mixer, in order that further signal processing may be applied if necessary. Some manufacturers have also added effects sends and returns.

It is also possible for some sound modules to have an audio input. This is not normally for sampling or recording purposes (unless the unit is a sampler) but allows an external audio signal to be mixed with the sound module's own output, as shown in Figure 3.6. In simple systems this approach can be used to mix the outputs of two sound modules together for example, obviating the need for an external mixer. In more advanced units it may also be possible to feed the external input through some of the sound module's internal processing.

3.1.6 Internal effects

Now that digital signal processing has become relatively cheap to implement, a number of sound modules and synths have their own effects units which include reverberation and other types of processing. In simple sound modules an effects unit will feed into the main stereo output of the module, as shown in Figure 3.7, and in more advanced units there may be a number of separate effects. Either from the front panel or under MIDI control it may be possible to control the send level from each voice to the effects unit, and the overall return level of the effects to the main stereo output can be adjusted to control the 'wet/dry' balance.

3.2 MIDI functions of synths and sound modules

The MIDI implementation for a particular synth or sound module should be described in the manual which accompanies it. A MIDI implementation chart such as the one shown in Figure 3.8 indicates which message types are received and transmitted, together with any comments relating to limitations or unusual features. Clearly sound modules without keyboards will be able to receive a lot more than they are able to transmit, and there may well be a number of 'hidden' functions only accessible through the use of system exclusive messages which will not appear on the implementation chart, although they should be documented separately. Functions such as note off velocity and polyphonic aftertouch are quite rare, although they may be found on certain modules.

It is quite common for a synthesiser to be able to accept certain data and act upon it, even if it cannot generate such data from its own controllers. The note range available under MIDI control compared with that available from the keyboard itself is a good example of this, since many devices will respond to note data over a full ten octave range yet still have only a limited keyboard. This approach can be used by a manufacturer who wishes to make a cheaper synthesiser which omits the expensive physical sensors for such things as velocity and aftertouch, while retaining these functions in software for use under MIDI control.

Synths conforming to the General MIDI specification described below must conform to certain basic guidelines concerning their MIDI implementation and the structure of their sound generators.

```
YAMAHA  [ Tone Generator ]
        Model  TG100   MIDI Implementation Chart   Version : 1.00
+----------------------------------------------------------------------+
:               : Transmitted  :   Recognized   :     Remarks         :
:      Function ... :          :                :                     :
:-------------------+-----------------+-----------------+--------------:
:Basic   Default : x           : 1 - 16          : memorized          :
:Channel Changed : x           : 1 - 16          :                    :
:-------------------+-----------------+-----------------+--------------:
:        Default : x           : 3               :                    :
:Mode    Messages: x           : 3,4(m = 1)  *2  :                    :
:        Altered : *************: x               :                    :
:-------------------+-----------------+-----------------+--------------:
:Note            : x           : 0 - 127         :                    :
:Number : True voice: *********** : 0 - 127      :                    :
:-------------------+-----------------+-----------------+--------------:
:Velocity Note ON  : x         : o   9nH,v=1-127 :                    :
:        Note OFF  : x         : x               :                    :
:-------------------+-----------------+-----------------+--------------:
:After   Key's   : x           : x               :                    :
:Touch   Ch's    : x           : o               :                    :
:-------------------+-----------------+-----------------+--------------:
:Pitch Bender    : x           : o  0-24 semi    :12bit resolution:   :
:-------------------+-----------------+-----------------+--------------:
:            0,32: x           : o MSB only      :Bank Select         :
:             1 : x            : o               :Modulation Wheel:   :
:             5 : x            : o               :Portamento Time     :
:          6,38: x            : o               :Data Entry          :
: Control     7 : x            : o           *1  :Volume              :
:            10 : x            : o               :Panpot              :
: Change     11 : x            : o           *1  :Expression          :
:            64 : x            : o               :Hold 1              :
:            65 : x            : o               :Portamento          :
:            91 : x            : o(Reverb)       :Effect Depth 1      :
:        100,101: x            : o               :RPN LSB,MSB         :
:           120 : x            : o               :All Sound Off       :
:           121 : x            : o               :Reset All Cntrls:   :
:               :             :                 :                    :
:-------------------+-----------------+-----------------+--------------:
:Prog            : x           : o 0-127     *1  :                    :
:Change : True # : *************:                :                    :
:-------------------+-----------------+-----------------+--------------:
:System Exclusive : o       *3 : o           *3  :                    :
:-------------------+-----------------+-----------------+--------------:
:System : Song Pos. : x        : x               :                    :
:       : Song Sel. : x        : x               :                    :
:Common : Tune    : x          : x               :                    :
:-------------------+-----------------+-----------------+--------------:
:System     :Clock  : x        : x               :                    :
:Real Time  :Commands: x       : x               :                    :
:-------------------+-----------------+-----------------+--------------:
:Aux   :Local ON/OFF : x       : x               :                    :
:      :All Notes OFF: x       : o(123-127)      :                    :
:Mes-  :Active Sense : x       : o               :                    :
:sages :Reset    : x           : x               :                    :
:-------------------+-----------------+-----------------+--------------:
:Notes: *1  ; receive if switch is on.                                :
:       *2  ; m is always treated as "1" regardless of its value.     :
:       *3  ; transmit/receive if exclusive switch is on.             :
:                                                                      :
:                                                                      :
+----------------------------------------------------------------------+
    Mode 1 : OMNI ON,  POLY   Mode 2 : OMNI ON,  MONO    o : Yes
    Mode 3 : OMNI OFF, POLY   Mode 4 : OMNI OFF, MONO    x : No
```

Figure 3.8 A typical MIDI implementation chart for a synthesiser sound module. (Yamaha TG100, with permission)

3.3 MIDI data buffers

All MIDI-controlled equipment uses some form of data buffering for received MIDI messages. Such buffering acts as a temporary store for messages which have arrived but not yet been processed, and allows for a certain prioritisation in the handling of received messages. Cheaper devices tend to have relatively small MIDI input buffers and these can overflow easily unless care is taken in the distribution of MIDI data around a large system (see Chapter 7). When a buffer overflows it will normally result in an error message displayed on the front panel of the device, indicating that some MIDI data is likely to have been lost. More advanced equipment can store more MIDI data in its input buffer, although this is not necessarily desirable because many messages that are transmitted over MIDI are intended for 'real-time' execution and one would not wish them to be delayed in a temporary buffer. A more useful solution would be to speed up the rate at which incoming messages can be processed, and more recent devices have considerably faster processors than older synths, leading to an improvement in this area. There is clearly a conflict here between the small buffer needed for real-time data, and the larger buffer required for system exclusive dumps.

3.4 Handling of velocity and aftertouch data

Synthesisers able to respond to note on velocity will use the value of this byte to control assigned functions within the sound generators. It is common for the user to be able to program the keyboard such that the velocity value affects certain parameters to a greater or lesser extent. For example, it might be decided that the 'brightness' of the sound should increase with greater key velocity, in which case it would be necessary to program the keyboard so that the envelope generator which affected the brightness was subject to control by the velocity value. This would usually mean that the maximum effect of the envelope generator would be limited by the velocity value, such that it could only reach its full programmed effect (that which it would give if not subject to velocity control) if the velocity was also maximum. The exact law of this relationship is up to the manufacturer, and may be used to simulate different types of 'keyboard touch'. A device may offer a number of laws or curves relating changes in velocity to changes in the control value, or the received velocity value may be used to scale the preset parameter rather than replace it.

Another common application of velocity value is to control the amplitude envelope of a particular sound, such that the output volume depends on how hard the key is hit. In many synthesiser systems which use multiple interacting digital oscillators, these velocity-sensitive effects can all be achieved by applying velocity control to the envelope generator of one or more of the oscillators, as indicated earlier in this chapter.

Note off velocity is not implemented in many keyboards, and most musicians are not used to thinking about what they do as they release a key, but this parameter can be used to control such factors as the release time of the note or the duration of a reverberation effect.

Aftertouch is an effect which is controlled by the amount of pressure applied at the bottom of a key's travel. It is often used in synthesisers to control the application of low frequency modulation (tremolo or vibrato) to a note. Sometimes aftertouch may be applied to other parameters, but this is less common. Generally, aftertouch is *channel* aftertouch because this is cheaper to implement, and does not produce as much MIDI data (see Chapter 2), but occasionally *polyphonic* aftertouch is implemented, which allows for greater control over expression based on the individual pressure on each key.

3.5 Channel modes

According to the MIDI standard an instrument should normally power up in omni on, polyphonic mode (see section 2.5.8), so that it accepts information on any channel, although in practice most devices power up in the mode to which they were last set. Thereafter the user may program it otherwise, either using MIDI control or on the front panel. It can often be useful to set a receiver to omni on mode in order to check whether any data at all is being received, for cases when problems arise. In most normal operations, though, a modern multi-timbral sound module would operate in omni off, polyphonic mode, receiving separate polyphonic data on each MIDI channel.

Channel mode 4 (omni off, mono), was used quite widely in the past as a means of getting more than one sound out of a synthesiser. Using mode 4, a 16 voice multi-timbral synthesiser could, under MIDI control, act as a moderately comprehensive 'orchestra' of sounds. Mode 4 has also been used widely with guitar controllers (see section 3.15.3). Tracks could be recorded on a sequencer by a master keyboard, each track set to play on a different MIDI channel, all channels to be received by one synthesiser. The 16 voices could be assigned to MIDI channels in whatever combination was desired, and it would not be mandatory to assign just one channel to a voice. One could, for example, assign channels 1 through 8 to a brass ensemble voice, channel 9 to a bass guitar voice, and the rest to solo voices. In this example it would depend on the sequencer in question as to exactly how the tracks were assigned to channels, and how the mode of the track was set, but some sequencers will allow a track's mode to be changed after the data has been recorded, such that a track recorded as 8 note polyphonic on channel 1 could be converted to play out in mode 4, with each of the eight notes on an adjacent channel from 1 to 8, to be played by the synthesiser configuration suggested above.

3.6 Pitch bend data

As pointed out in Chapter 2, pitch bend data is channel specific. On instruments where it is possible to receive on more than one channel, as indicated above, pitch bend data should act only on the notes sounded as a result of note messages on the same channel as itself. There are, though, certain instruments which treat the pitch bend as applying across the whole instrument, and this is normally undesirable.

3.7 Controller messages
3.7.1 General

The controller messages which begin with a status of &Bn, as listed in Table 2.4, turn up in various forms in synthesiser implementations. It should be noted that although there are standard definitions for many of these controller numbers it is often possible to remap them either within sequencer software or within sound modules themselves. 14 bit continuous controllers are rarely encountered for any parameter, and often only the MSbyte of the controller value (which uses the first 32 controller numbers) is sent and used. For most parameters the 128 increments which result are adequate resolution. The MIDI implementation chart for the device in question will indicate if any 14 bit controllers are in operation.

Controllers &07 (Volume) and &0A (Pan) are particularly useful with sound modules as a means of controlling the internal mixing of voices. These controllers work on a per channel basis, and are independent of any velocity control which may be related to note volume. There are two real-time system exclusive controllers which handle similar functions to these, but for the device as a whole rather than for individual voices or channels. The 'master volume' and 'master balance' controls are accessed using:

&[F0] [7F] [dev. ID] [04] [01 or 02] [data] [data] [F7]

where the sub-ID #1 of &04 represents a 'device control' message and sub-ID #2s of &01 or &02 select volume or balance respectively. The [data] values allow 14 bit resolution for the parameters concerned, transmitted LSB first. Balance is different to pan because pan sets the stereo positioning (the split in level between left and right) of a mono source, whereas balance sets the relative levels of the left and right channels of a stereo source (see Figure 3.9). Since a pan or balance control is used to shift the image either left or right from a centre detent position, the MIDI data values representing the setting are ranged either side of a mid-range value which

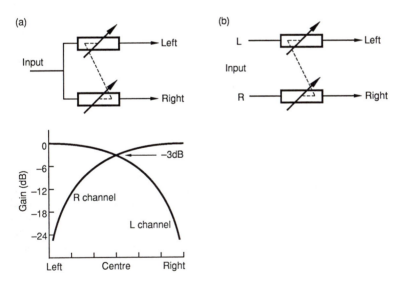

Figure 3.9 (a) A pan control takes a mono input and splits it two ways (left and right), the stereo position depending on the level difference between the two channels. The attenuation law of pan controls is designed to result in a smooth movement of the source across the stereo 'picture' between left and right, with no apparent rise or fall in overall level when the control is altered. A typical pan control gain law is shown below. (b) A balance control simply adjusts the relative level between the two channels of a stereo signal so as to shift the entire stereo image either left or right

Table 3.1 An example of pan controller mapping

Pan position	MIDI pan controller value
–7 (left)	0–7
–6	8–15
–5	16–23
–4	24–31
–3	32–39
–2	40–47
–1	48–55
0 (centre)	56–63
+1	64–71
+2	72–79
+3	80–87
+4	88–95
+5	96–103
+6	104–111
+7 (right)	112–127

corresponds to the centre detent. The channel pan controller is thus normally centred at a data value of 63 (and sometimes over a range of values just below this if the pan has only a limited number of steps), assuming that only a single 7 bit controller value is sent. For example, the E-mu Proteus sound modules use a pan law which has steps from –7 (corresponding to fully left) to +7 (corresponding to fully right), as shown in Table 3.1.

Some manufacturers have developed alternative means of expressive control for synthesisers such as the 'breath controller', which is a device which responds to the blowing effort applied by the mouth of the player. It was intended to allow wind players to have more control over expression in performance. Plugged into the synthesiser, it can be applied to various envelope generator or modulator parameters to affect the sound. The breath controller also has its own MIDI controller number.

There is also a recent portamento controller (&54) which defines a note number from which the next note should slide. It is normally transmitted between two note on messages to create an automatic legato portamento effect between two notes.

3.7.2 Effects and sound controllers

Recent devices may also respond to the 'effects' and 'sound' controllers. These controller ranges have been set aside as a form of general purpose control over certain aspects of the built-in effects and sound quality of a device. How they are applied will depend considerably on the architecture of the sound module and the method of synthesis used, but they give some means by which a manufacturer can provide a more abstracted form of control over the sound without the user needing to know precisely which voice parameters to alter. In this way, a user who is not prepared to get into the increasingly complicated world of voice programming can modify sounds to some extent.

The effects controllers occupy five controller numbers from &5B to &5F, and are defined as Effects Depths 1–5. The default names for the effects to be controlled by these messages are respectively 'External Effects Depth', 'Tremolo Depth', 'Chorus Depth', 'Celeste (Detune) Depth and 'Phaser Depth', although these definitions are open to interpretation and change by manufacturers.

Table 3.2 Sound controller functions (byte 2 of status &Bn)

MIDI controller number	Function (default)
&46	Sound variation
&47	Timbre/harmonic content
&48	Release time
&49	Attack time
&4A	Brightness
&4B–4F	No default

There are also ten sound controllers which occupy controller numbers from &46 to &4F. Again these are user- or manufacturer-definable, but there are currently defaults for the first five (listed in Table 3.2). They are principally intended as real-time controllers to be used during performance, rather than as a means of editing internal voice patches (the RPCs and NRPCs can be used for this as described below).

The sound variation controller is interesting because it is designed to allow the selection of one of a number of variants on a basic sound, depending on the data value which follows the controller number. For example, a piano sound might have variants of 'honky tonk', 'soft pedal', 'lid open' and 'lid closed'. The data value in the message is not intended to act as a continuous controller for certain voice parameters, rather the different data values possible in the message are intended to be used to select certain pre-programmed variations on the voice patch. If there are less than the 128 possible variants on the voice then the variants should be spread evenly over the number range so that there is an equal number range between them.

The timbre and brightness controllers can be used to alter the spectral content of the sound. As described earlier in this chapter, the means by which sounds are generated varies and thus the precise parameter to which these messages would be applied cannot easily be defined, but the timbre controller is intended to be used specifically for altering the harmonic content of a sound, whilst the brightness controller is designed to control its high frequency content.

The envelope controllers can be used to modify the attack and release times of certain envelope generators within a synthesiser. Data values less than &40 attached to these messages should result in progressively shorter times, whilst values greater than &40 should result in progressively longer times.

3.7.3 Registered and non-registered parameter numbers

The MIDI standard was extended a few years ago to allow for the control of individual internal parameters of sound generators by using a specific control change message. This meant, for example, that any aspect of a voice, such as the velocity sensitivity of an envelope generator, could be assigned a parameter number which could then be accessed over MIDI and its setting changed, making external editing of voices much easier. Parameter controllers are a subset of the control change message group, and they are divided into the registered and non-registered numbers (RPNs and NRPNs). RPNs are intended to apply universally, and to be registered with the MIDI Manufacturers' Association (MMA), whilst NRPNs may be manufacturer specific.

Parameter controllers operate by specifying the address of the parameter to be modified, followed by a control change message to increment or decrement the

Table 3.3 Some examples of RPC definitions

RPC number (hex)	Parameter
00 00	Pitch bend sensitivity
00 01	Fine tuning
00 02	Coarse tuning
00 03	Tuning program select
00 04	Tuning bank select
(7F 7F	Cancels RPN or NRPN (usually follows Message 3)

setting concerned. It is also possible to use the data entry slider controller to alter the setting of the parameter. The address of the parameter is set in two stages, with an MSbyte and then an LSbyte message, so as to allow for 16 384 possible parameter addresses. The controller numbers &62 and &63 are used to set the LS- and MSbytes respectively of an NRPN, whilst &64 and &65 are used to address RPNs.

The sequence of messages required to modify a parameter is as follows:
Message 1

&[Bn] [62 or 64] [LSB]

Message 2

&[Bn] [63 or 65] [MSB]

Message 3

&[Bn] [60 or 61] [7F] or &[Bn] [06] [DATA] [38] [DATA]

Message 3 represents either data increment (&60) or decrement (&61), or a 14 bit data entry slider control change with MSbyte (&06) and LSbyte (&38) parts (assuming running status). If the control has not moved very far, it is possible that only the MSbyte message need be sent.

Only five parameter numbers are currently registered as RPNs, as shown in Table 3.3, but more may be added at any time and readers are advised to check the most recent revisions of the MIDI standard.

3.8 Voice selection

3.8.1 Program change

The program change message was adequate for a number of years as a means of selecting one of a number of stored voice patches on a synthesiser. Program change on its own allows for up to 128 different voices to be selected, and its basic application was described in section 2.5.9. (Remember that there is often an offset of one between the actual MIDI program change data byte and the number of the program to be selected.) A synthesiser or sound module may allow a program change map to be set up in order that the user may decide which voice is selected on receipt of a particular message. This can be particularly useful when the module has more than 128 voices available, but no other means of selecting voice banks. A number of different program change maps could be stored, perhaps to be switched between under system exclusive control.

3.8.2 Bank select

Modern sound modules tend to have very large patch memories – often too large to be adequately addressed by 128 program change messages. Although some older synthesisers used various odd ways of providing access to further banks of voices, most recent modules have now implemented the standard 'bank select' approach. In basic terms, 'bank select' is a means of extending the number of voices which may be addressed by preceding a standard program change message with a message to define the bank from which that program is to be recalled. It uses a 14 bit control change message, with controller numbers &00 and &20, to form a 14 bit bank address, allowing 16 384 banks to be addressed. The bank number is followed directly by a program change message, thus creating the following general message:

&[Bn] [00] [MSbyte (of bank)]

&[Bn] [20] [LSbyte]

&[Cn] [Program number]

If a receiver does not understand bank select messages then it will simply ignore them and the result would be the selection of only the program number defined in the last part of the message. (There is some confusion over the sending of just MSbytes, and this has yet to be resolved, see section 3.8.4.)

3.8.3 General MIDI

One of the problems with MIDI sound generators is that although voice patches may be selected using MIDI program change commands, there is no guarantee that a particular program change number will recall a particular voice on more than one instrument. In other words, program change 3 may correspond to 'alto sax' on one instrument and 'grand piano' on another. This makes it difficult to exchange songs between systems with any hope of the replay sounding the same as intended by the composer. For this reason General MIDI was specified as a means of standardising aspects such as this, so that songs could be exchanged more easily between systems, and so that voices would be mapped in a fashion which ensured that at least an approximation to the same sound would be produced by each instrument for a particular program number. Currently, General MIDI is specified at Level 1, although there are proposals to extend the concept to further levels.

General MIDI specifies a number of other things as well as standard sounds. For example, it specifies a minimum degree of polyphony, and requires that a sound generator should be able to receive MIDI data on all 16 channels simultaneously and polyphonically, with a different voice on each channel. There is also a requirement that the sound generator should support percussion sounds in the form of drum kits, so that a General MIDI sound module is capable of acting as a complete 'band in a box'.

Dynamic voice allocation is considered to be the norm in GM sound modules, with a requirement either for at least 24 dynamically allocated voices in total, or 16 for melody and 8 for percussion. Voices should all be velocity sensitive, and should respond at least to the controller messages 1, 7, 10, 11, 64, 121 and 123 (decimal), RPNs 0, 1 and 2 (see above), pitch bend and channel aftertouch. In order to ensure compatibility between sequences that are replayed on GM modules, percussion

Table 3.4 General MIDI program number ranges (except channel 10)

Program change (decimal)	Sound type
0–7	Piano
8–15	Chromatic percussion
16–23	Organ
24–31	Guitar
32–39	Bass
40–47	Strings
48–55	Ensemble
56–63	Brass
64–71	Reed
72–79	Pipe
80–87	Synth lead
88–95	Synth pad
96–103	Synth effects
104–111	Ethnic
112–119	Percussive
121–128	Sound effects

sounds are always allocated to MIDI channel 10. Program change numbers are mapped to specific voice names, with ranges of numbers allocated to certain types of sounds, as shown in Table 3.4. Precise voice names may be found in the GM documentation. Channel 10, the percussion channel, has a defined set of note numbers on which particular sounds are to occur, so that the composer may know for example that key 39 will always be a 'hand clap'.

General MIDI sound modules may operate in modes other than GM, where voice allocations may be different, and there are two universal non-realtime SysEx messages used to turn GM on or off. These are:

&[F0] [7E] [dev. ID] [09] [01] [F7]

to turn GM on, and:

&[F0] [7E] [dev. ID] [09] [02] [F7]

to turn it off.

There is some disagreement over the definition of 'voice', as in '24 dynamically allocated voices' – the requirement which dictates the degree of polyphony that should be supplied by a GM module. The spirit of the GM specification suggests that 24 notes should be capable of sounding simultaneously, but some modules combine sound generators to create composite voices, thereby reducing the degree of note polyphony.

3.8.4 Roland GS

The manufacturer Roland has developed an extension/variation on GM, called 'GS'. GS extends the number of available voices above 128 by using a form of bank select message to switch in 'variation' banks on the fundamental sounds of GM. The basic GM voices are called 'Capital Tones' and are selected by bank 0, whilst the alternatives are called 'Variation Tones'. There may also be 'Sub-capital Tones' which have a less clear relationship to the original GM sounds. Program change numbers sent whilst in the variation banks still select the same *type* of sound as in bank 0 (e.g.: a violin sound will still be a violin sound), but with an altered timbre. In something of a deviation from the MIDI standard, GS banks are selected using only the MSbyte of the bank select message (i.e.: controller 0).

3.9 Tuning

Conventional equal-tempered tuning is the norm in western musical environments, but there may be cases when alternative tuning standards are required in order to conform to other temperaments or to non-western musical styles. Many devices now have the capability to store a number of alternative tuning maps, or to be retuned 'on the fly'. A number of manufacturer-specific methods have been used in the past (prior to the MIDI Tuning Standard), being SysEx messages preceded by the relevant manufacturer ID, but it is likely that the MIDI Tuning Standard will form the basis for communicating information about alternative tunings in all future devices.

3.9.1 Musical basis of the tuning standard

The tuning standard has to assume that any note on a synth can be tuned over the entire range 8.1758 Hz to 13289.73 Hz. It then allows individual notes' tuning to be adjusted in fractions of a semitone above a conventional MIDI note's pitch (which would be based on the equal temperament convention). A semitone is divided into 100 cents. A cent is one hundredth of a semitone, and as such does not represent a constant frequency increment in hertz but represents a proportion of the frequency of the note concerned. As the pitch of the basic note rises, so the frequency increment represented by a cent also increases. Two MIDI data bytes are used to indicate the fraction of a semitone above the basic note pitch, so the maximum resolution possible is 100 cents/2^{14} which equals 0.0061 cents.

Tuning of individual notes is represented by three MIDI messages in total. The first specifies a numbered semitone in the MIDI note range on which the fractional tuning is to be based (the same as the MIDI note number in a note on message) and the second and third form a pair containing a 14 bit number (the MSB of each is 0), transmitted MSB first. This 14 bit number is used as described in the previous paragraph, with each increment representing a change of 0.0061 cents upwards in pitch from the basic semitone number (see Figure 3.10). A synth which is not capable of tuning to the accuracy contained in the message should tune to the nearest possible value, but it is recommended that it *stores* the full resolution tuning value

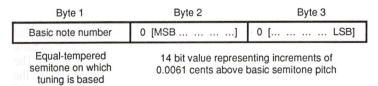

Byte 1	Byte 2	Byte 3
Basic note number	0 [MSB ... ...]	0 [... ... ... LSB]

Equal-tempered   14 bit value representing increments of
semitone on which   0.0061 cents above basic semitone pitch
tuning is based

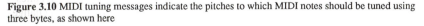

Figure 3.10 MIDI tuning messages indicate the pitches to which MIDI notes should be tuned using three bytes, as shown here

in tuning memories, in case data is to be transmitted to other devices which are capable of full resolution. The frequency value of &[7F] [7F] [7F] is reserved to indicate no change to the tuning of a particular note.

3.9.2 MIDI tuning messages

A number of MIDI messages are associated with tuning. These break down into bulk dumps of tuning data (to retune a complete instrument), single note retuning messages and the selection of prestored tuning programs and banks of programs. The only one of these which is currently a real-time message is the single note retuning.

A device may request a bulk tuning dump from another using the general SysEx non-realtime form:

&[F0] [7E] [dev. ID] [08] [00] [tt] [F7]

where the sub-ID #1 of &08 indicates a MIDI tuning standard message and the sub-ID #2 of &00 indicates a bulk dump request. &[tt] defines the tuning program which is being requested. Such a request should result in the transmission of a bulk dump if such a tuning program exists, and the dump should take the form:

&[F0] [7E] [dev. ID] [08] [01] [tt] [tuning name] ... ... [tuning data] ... ...
... [LL] [F7]

where [tuning name] is 16 bytes to name the tuning program (each byte holds a 7 bit ASCII character), and [tuning data] consists of 128 groups of 3 bytes to define the tuning of each note, in the format described in the previous section. &LL is a checksum byte.

A single note may be retuned using the SysEx realtime message:

&[F0] [7F] [dev. ID] [08] [02] [tt] [ll] ([kk] [tuning data]) ... ... [F7]

where &[ll] indicates the number of notes to be retuned, followed by that number of groups of tuning data. Each group of tuning data is preceded by &[kk] which defines the note to be retuned.

Users may select stored tuning programs and banks of tuning programs using the RPN messages shown in Table 3.3.

3.10 Voice data dumps

It is normally possible to program a synthesiser's sounds oneself, and a facility should be provided for the storage of these sound programs, 'voices' or 'patches'. Different manufacturers have adopted varying methods for such storage, ranging from RAM cartridges to disk drives, and it may be possible to dump the data relating to a voice program using a system exclusive message over MIDI. Such a dump could be used either to transfer voice data between two synths of a similar type, or to dump voice data to a computer for storage or editing.

In the case of the former, a full keyboard version of a particular make of synthesiser could be used to program an expander version which might have few controls of its own. It is often advantageous to be able to transfer voice data to a computer. One particularly persuasive reason is an economical one, since a disk can hold many thousand voices whereas the internal memory of the synth might only store a few. The plug-in RAM cards or cartridges sold by synth manufacturers are often expensive, and computer mass storage is relatively cheap. Software packages for voice editing are also available, which interpret the system exclusive voice data graphically, allowing the user to edit the sound in an easier manner than is possible with the limited display on the synthesiser. Data is passed back and forth over MIDI between the synthesiser and the computer, using system exclusive messages, to update both the display and the sound.

There is currently no universal format for voice parameter data, since the methods of synthesis used in different sound generators vary so much. For this reason the voice dump messages tend to be manufacturer specific, but voice editors usually allow the installation of interpreters for a wide range of commercial synthesisers, with new modules being issued to deal with newly released products. Voice dumps will often be arranged so that there is one system exclusive message which handles individual voice dumps, and another that handles dumps of a whole bank. This is useful as it means that the editing of one voice need not result in the loss of all the other stored voices in the synthesiser. It is to be hoped that more use will be made in future of the RPN and NRPN controller messages (see section 3.7.3) for the editing of voice parameters, which would help in making voice editing software more able to handle multiple devices without each one needing a separate set of SysEx messages.

3.11 Introduction to samplers

This section will introduce the basic principles involved in the sampling of music signals, as used in MIDI-based samplers. It is intended to provide a background to the understanding of the various MIDI control protocols used with samplers, explaining the terminology used.

3.11.1 What is a sampler?

A sampler, unlike a synthesiser, stores and replays 'real' sounds in a digital form. It is used for 'recording' audio information sourced from either a line input or a microphone, and allows the stored sound to be edited, looped, and otherwise processed, before using a software program to control the replay of the sample under MIDI control. Although a sampler is used to store and replay digital audio, it differs conceptually from a conventional digital audio recorder in a number of ways. Stored sounds are replayed instantaneously from memory, triggered by MIDI or keyboard messages, and the architecture of a sampler is normally arranged so as to allow pitch transposition of a sample as well as a wide range of musical processing options. A sampler is a real-time 'musical instrument' normally used for storing and processing short self-contained 'samples', whereas a digital audio recorder is designed for the storage and editing of complete programmes.

That said, the basic principles of digital audio are exactly the same for both of the above devices – it is just that the application of the technology is slightly different. The term 'sample' has come to mean a stored 'sound' in music applications, but this is a misuse of the term, as discussed below. A music sampler consists of a certain amount of RAM storage, normally some form of mass storage, a microcomputer to control operations, a user interface, a signal processing block and a collection of A/D and D/A convertors connected to audio inputs and outputs. This is illustrated in Figure 3.11.

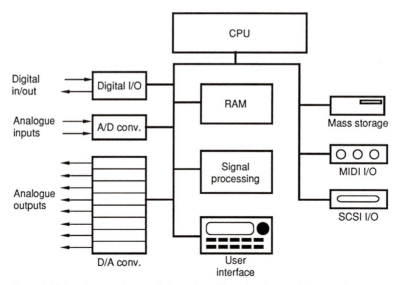

Figure 3.11 A typical MIDI-controlled sampler will consist of most of the operational blocks shown

Modern samplers and synthesisers often have many features in common, making the dividing line between the two less clear as time progresses. For example, synthesisers often store sampled sounds and use them in creating new ones, whereas samplers often incorporate synthesis features for altering the timbre of stored sounds. Whether the device is called a sampler or a synthesiser is really dependent on the emphasis which the manufacturer chooses to give to certain features.

3.11.2 Principles of digital sampling

Sound is stored in a sampler in a digital form. Because of this any analogue audio signals connected to the input must be converted into binary data, and the binary data must be converted back into analogue form at the audio outputs. Some samplers are also able to handle inputs and outputs in a digital form for direct connection to other digital studio equipment. (For further details of digital interface formats and practice the reader is referred to *The Digital Interface Handbook* by Rumsey and Watkinson, as detailed in Appendix 2.)

The process of A/D conversion turns the electrical waveform from a microphone or line input into a series of binary numbers, each of which represents the amplitude of the signal at a point in time. The analogue audio signal (a time-varying electrical voltage) is passed through convertor where it is transformed from a continuously varying voltage into a series of 'samples'. The samples can be considered as rather like instantaneous 'still frames' of the audio signal which together and in sequence form a representation of the continuous waveform, rather as the still frames which make up a movie film give the impression of a continuously moving picture when played in quick succession (see Figure 3.12). Each sample amplitude is then converted into a binary number, in a process known as quantisation. On replay the

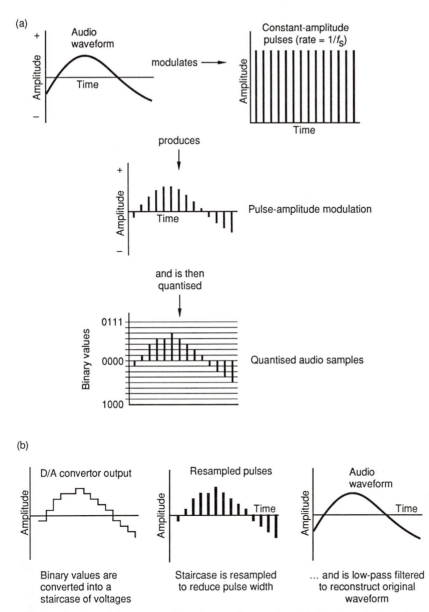

Figure 3.12 (a) In an A/D convertor, the analogue audio wave is sampled as shown. The sample amplitudes represent the instantaneous amplitude of the audio signal at regularly spaced points in time. These samples are then quantised to give them binary values (4 bit scale shown). (b) In the D/A conversion process, the binary values are turned back into an analogue wave

data is fed to a D/A convertor which turns the numerical data back into a time-continuous analogue audio signal.

In order to represent the fine detail of the signal it is necessary to take a large number of samples per second, and the mathematical sampling theorem proposed by Shannon indicates that at least two samples must be taken per audio cycle if the necessary information about the signal is to be conveyed. It can be seen from Figure 3.13 that if less than two samples are taken per cycle then it becomes possible to reconstruct a wave other than that originally sampled, and this is the phenomenon known as aliasing. The aliasing phenomenon can be seen in the case of the well-known 'spoked-wheel' effect on films, since moving pictures are also an example of a sampled signal. In film, still pictures (visual samples) are normally taken at a rate of 24 per second. If a rotating wheel with a marker on it is filmed it will appear to move round in a forward direction as long as the rate of rotation is much slower than the rate of the still photographs, but as its rotation rate increases it will appear to slow down, stop, and then appear to start moving backwards. The virtual impression of backwards motion gets faster as the rate of rotation of the wheel gets faster, and this backwards motion is the aliased result of sampling at too low a rate. If audio signals are allowed to alias in digital recording one hears the audible equivalent of the backwards-rotating wheel – that is, sound components in the audible spectrum which were not there in the first place, moving downwards in frequency as the original frequency of the signal increases.

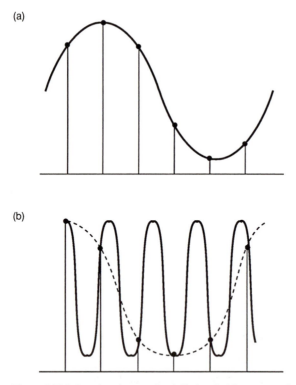

(a)

(b)

Figure 3.13 A time-domain example of aliasing. In (a) many samples are taken per cycle of the wave, and the reconstruction low-pass filter recreates the original wave on replay. In (b), less than two samples are taken per cycle, making it possible for another lower-frequency wave to be reconstructed from the samples

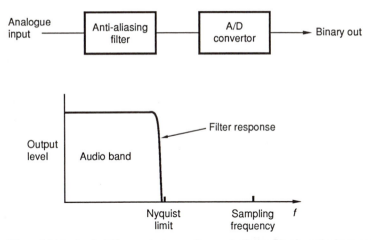

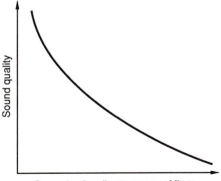

Figure 3.14 In simple A/D convertors an analogue anti-aliasing filter is used prior to conversion, which removes input signals with a frequency above the Nyquist limit. More advanced convertors employ a process known as oversampling, which removes the need for analogue filters of this type by effectively performing the operation in the digital domain

Figure 3.15 High sound quality requires a high sample frequency and a large number of bits per sample. There is therefore a direct tradeoff between memory usage and sound quality in a digital audio sampler

The rate at which samples are taken is known as the sampling frequency, and this is directly related to the frequency response of the system, since only frequencies up to half the sampling frequency may be represented without aliasing. Therefore in order to represent audio frequencies up to 20 kHz (the upper limit of hearing) it is necessary to have a sampling frequency of at least 40 kHz. In basic convertors, a filter, known as an anti-aliasing filter, is used prior to sampling whose job is to remove all input signals above half the sampling frequency (sometimes called the Nyquist frequency), as shown in Figure 3.14. In practice, the sampling rate of full-range audio systems such as the compact disc is slightly higher than that required for minimal adherence to theory, and has been set at 44.1 kHz. This allows for filters which do not have such a sharp cut-off above the Nyquist frequency, and this in turn has benefits for sound quality since steep filters tend to 'ring' and affect the sound at high frequencies.

In MIDI-based samplers it may be possible to choose the sampling frequency to suit the bandwidth of the signal to be stored. It may also be possible to 'resample' stored sounds at a lower rate once they are in memory, resulting in a reduction in bandwidth. A lower sampling rate results in the generation of a smaller amount of data per second, and may help in the optimisation of memory usage, although too low a sampling rate will restrict the bandwidth of a stored sound and may make it appear 'woolly' or 'muffled'. Sound quality must be traded off against memory usage, as shown in Figure 3.15.

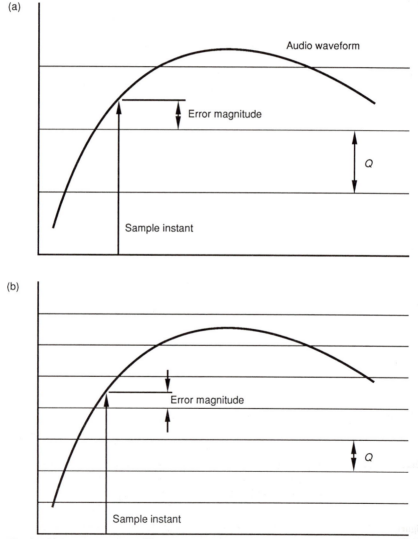

Figure 3.16 Quantising error. At (a) there are not many quantising levels and so the maximum error of ±0.5 Q is larger than at (b) where there are more quantising levels

The number of bits used in the binary words which represent each sample value define the quantising accuracy. From a sound quality point of view, the more bits the higher the sound quality. The number of bits determines the number of quantising levels available to represent the amplitude of the signal, and the more quantising levels there are the smaller will be the quantising error, as shown in Figure 3.16. In audio terms, a smaller quantising error results in lower distortion and noise. An 8 bit digital audio system has 2^8 (256) quantising levels and will sound only passable with a signal-to-noise ratio (SNR) of around 48 dB (early MIDI samplers used 8 bits), whereas more recent systems using 16 bits have 2^{16} (65536) quantising levels and will have very low noise and distortion (SNR is approximately 96 dB). 16 bits is the standard used in the compact disc system. The disadvantage of using a larger number of bits is that the memory requirement increases, and thus the same tradeoff exists between sound quality and memory usage as was shown in Figure 3.15.

To summarise, then, ultimate sound quality in a sampler is influenced by the sampling rate and the number of bits of resolution. Sampling rate affects the frequency response or bandwidth of the signal, and resolution affects the signal-to-noise ratio and the amount of distortion produced. Most modern MIDI samplers use 16 bit resolution (not user selectable), and may offer a selection of sampling rates from about 50 kHz downwards. It is common for recent professional samplers to operate at the CD rate of 44.1 kHz.

3.11.3 Sample storage and replay

When a sound is sampled and stored in a sampler's memory it will normally be stored at full length and full resolution, tagged with a particular MIDI note number specified by the user and information about the sampling rate originally used during A/D conversion. The note number indicates the original pitch of the sampled sound and is used as a reference for any subsequent transposition of the sound. In order to make the sound usable in a musical or sound effects environment it will normally require a certain amount of editing, and a program will be devised by the user to control the replay of the sample. Depending on the sampler it may store either mono or stereo samples. An example is given below of the process as might be applied to the sampling of an organ note.

The editing process involves 'topping and tailing' the sound to remove unwanted noises or silence at both ends (see Figure 3.17). This leaves only the wanted information and frees memory which would otherwise be wasted. For sounds which

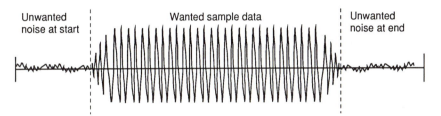

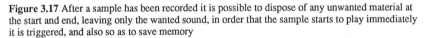

Figure 3.17 After a sample has been recorded it is possible to dispose of any unwanted material at the start and end, leaving only the wanted sound, in order that the sample starts to play immediately it is triggered, and also so as to save memory

would normally be sustained on replay it is normal then to introduce a looped section into the sample's program. This is because, for example, the sampled note might only have lasted one second during recording, whereas a musician might wish to sustain it for longer on replay. Without the loop the sample would simply be played through once and then stop, no matter how long the key was held. Once these editing processes have been performed the remaining audio data which is not part of the edited sample can be discarded if required, which normally frees up considerable memory since only a short portion of the original note will have been used in the loop. The remaining sample data can then be stored permanently on disk for later recall, along with the information about original pitch, sampling rate and looping information. Samplers may be equipped with a simple floppy disk drive, or they may be provided with a SCSI interface in order that external mass storage may be connected. A SCSI interface also allows for the use of a CD-ROM drive, making it possible to load sounds from CD-ROMs on which commercial libraries of sounds are often distributed.

A loop is defined either automatically by the sampler software, or manually by the user. The loop start and end points are chosen so as (a) to make the replay sound realistic by taking the most suitable portion of the original sound, and (b) to minimise the audible effect of joining one part of the waveform to another. As shown in Figure 3.18, the loop start point will be a certain distance into the sound, and the end point will be a certain number of samples later. On replay, the sampler would then play from the 'note on' point to the 'loop end' point, and then cycle round and round the loop until a 'note off' message was received. The audible effect of the join from loop end to loop start is minimised by ensuring that the audio waveform is edited either at a 'zero-crossing' point or at another point which ensures continuity of the cycle, so that no sharp change in sample amplitude occurs (which would result in a click or 'glitch'), as shown in Figure 3.19.

A number of software programs are available for editing and looping samples, and some of these have advanced digital signal processing (DSP) features which

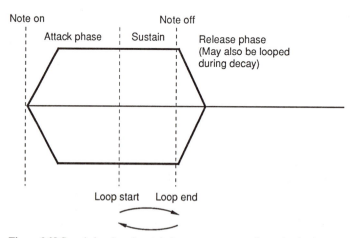

Figure 3.18 Sample looping. A sustain loop starts at a user-determined point during the sound and ends at a point which can be joined successfully back to the start of the loop. The loop plays as long as the note is held on

(a)

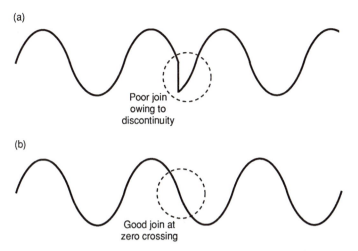

Poor join
owing to
discontinuity

(b)

Good join at
zero crossing

Figure 3.19 Sample loop points should be chosen so as to avoid discontinuities in the waveform, such as that shown at (a). A zero crossing point, as shown at (b), is a good place to join the waveform

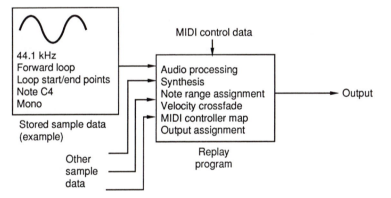

MIDI control data

44.1 kHz
Forward loop
Loop start/end points
Note C4
Mono

Stored sample data
(example)

Other
sample
data

Audio processing
Synthesis
Note range assignment
Velocity crossfade
MIDI controller map
Output assignment

Replay
program

Output

Figure 3.20 Sample data and the program which is used to replay the samples are separate entities

allow crossfading and other blending techniques at loop joins. Such techniques are particularly useful when editing stereo samples, since it is difficult to find a loop point which works ideally for both left and right channels simultaneously. It may be possible to introduce more than one loop into a sample program, in order, for example, to have a loop during the release phase of a note, as well as or instead of during the sustain portion.

The program used to replay the sample should be clearly distinguished from the sample data and its 'vital statistics'. The replay program determines which MIDI notes will trigger which samples, how the replay will respond to MIDI control information such as velocity, and whether any additional signal processing is applied to the sound. The illustration in Figure 3.20 shows this graphically. The replay program is normally stored separately to the sample data, and may be designed to organise the replay of more than one sample. Effectively the replay program defines the 'instrument' configuration, and it is possible to have a number of programs for one collection of samples if necessary. The following list is an example of the parameters which are commonly controlled by the replay program, many of which are related to MIDI control messages:

Note range assignment of samples; crossfades between ranges; velocity crossfades; envelope characteristics; filtering characteristics; scaling of control information; detuning; LFO modulation; velocity sensitivity of various parameters; panning between left and right outputs (for mono samples); output assignment.

The replay program will assign certain sounds to certain note ranges, as discussed further below. Although a sound might have been sampled at only one pitch it may be transposed over as wide a range as required by the replay program. In basic samplers, transposition on replay can be achieved by varying the replay sampling rate so that the stored data representing the original sound is read out either faster or slower than normal, resulting in a proportional change of pitch. The reconstruction filter after D/A conversion would have to track the sampling frequency in such a case. By comparing the original pitch of the sound to the pitch required when a note message is received the sampler software determines the degree of transposition required. For example, if the original pitch of the sample was noted as G3 and replay was required at G4, the sampling rate would be increased by a factor of two over the nominal rate of the sound, thus doubling the frequency of the replayed note. (An octave change in pitch is achieved by a doubling in frequency.) In recent samplers, DSP techniques are used to pitch-shift samples, involving the use of pitch-shifting algorithms based on digital decimation/interpolation operations to calculate the new sample values required at the pitch-shifted frequency.

Experimentation is required to determine the range over which a sample can successfully be transposed, and often only a limited range is possible before the timbre begins to sound unusual. (The spectral content of real instruments, as well as other characteristics, changes a lot with pitch, and thus it is not normally possible to take only a single sample of an instrument and transpose it over the whole pitch range.)

The foregoing information has been concerned with samplers that both record and replay sounds, but there also exist replay-only sample modules in which banks of prestored sounds are kept in ROM. These modules often include additional sound processing facilities so that the user can modify the stored voices, but they do not allow the original sample waveform to be changed.

3.11.4 Audio routing and processing

Depending on the sampler, audio for each group of notes may be replayed either to a simple stereo output or to one of a number of individual outputs. In the former case the replay program or MIDI panning controller data determines the panning of audio between left and right, and in the latter the replay program normally assigns particular note ranges to particular outputs. The latter approach is useful because it allows individual sounds to be connected to separate inputs of an external mixer, which then makes it possible for them to be processed individually.

Replayed sample data may be passed through stages of audio processing within the sampler before reaching the audio outputs, and a number of these possibilities are illustrated in Figure 3.21. Many of the audio processing features found in

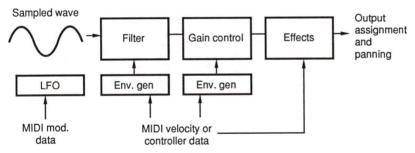

Figure 3.21 The sampled wave may be subjected to synth-like signal processing before the output

synthesisers can be included at this point, and this provides a means for altering the envelope and timbral characteristics of sampled sounds, as well as possibly the addition of effects such as reverberation. Such modification of the sound may indeed prove necessary, because the process of editing and looping the sample often destroys a lot of the natural variation in the sound's quality as a note is sustained. (Only a short portion of the original note will normally have been retained for the loop.) The various modulators and other effects can be used to restore some 'character' to a neutral sound in such cases.

As in synthesisers, MIDI velocity information may be used to influence changes in timbre or volume, based on such processing of the sampled sound. Velocity value applied to the filter cutoff frequency for example, could be used to make the sound brighter as the velocity value increased.

3.12 MIDI note assignment in samplers

A sampler's memory may be divided between a number of different stored sounds. It is possible that these sounds could be entirely unrelated (perhaps a single drum, an animal noise and a piano note), or that they might have some relationship to each other (perhaps a number of drums in a kit, or a selection of notes from a grand piano). The method by which samples are assigned to MIDI notes is likely to differ between these examples, and is defined by the replay program (see above).

3.12.1 Basic note assignment

The most common approach when assigning note numbers to samples is to program the sampler with the range of MIDI note numbers over which a certain sample should be sounded. Akai, one of the most popular sampler manufacturers, calls these 'keygroups'. It may be that this 'range' is only one note, in which case the sample in question would be triggered only on receipt of that note number, but in the case of a range of notes the sample would be played on receipt of any note in the range. In the latter case transposition would be required, depending on the relationship between the note number received and the original note number given to the sample (see above). A couple of examples highlight the difference in approach, as shown in Figure 3.22.

In the first example, illustrating a possible approach to note assignment for a collection of drum kit sounds, most samples are assigned to only one note number,

(a)

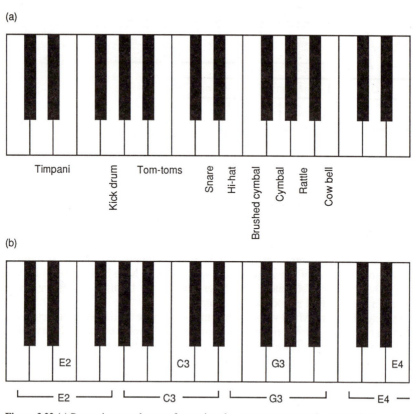

Figure 3.22 (a) Percussion samples are often assigned to one note per sample, except for tuned percussion which sometimes covers a range of notes. (b) Organ samples could be transposed over a range of notes, centred on the original pitch of the sample

although it is possible for tuned drum sounds such as tom-toms to be assigned over a range in order to give the impression of 'tuned toms'. Each MIDI note message received would replay the particular percussion sound assigned to that note number in this example.

In the second example, illustrating a suggested approach to note assignment for an organ, notes were originally sampled every musical fifth across the organ's note range. The replay program has been designed so that each of these samples is assigned to a note range of a fifth, centred on the original pitch of each sample, resulting in a maximum transposition of a third up or down. Ideally, of course, every note would have been sampled and assigned to an individual note number on replay, but unless the sampler has copious amounts of memory and a very flexible replay program this is not normally practical. This, then, is a suitable compromise.

3.12.2 Positional crossfades

In further pursuit of sonic accuracy, some samplers provide the facility for introducing a crossfade between note ranges. This is used where an abrupt change

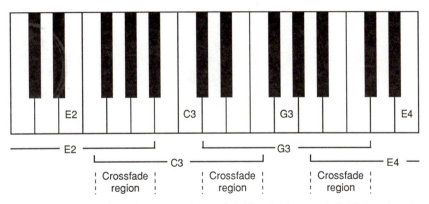

Figure 3.23 Overlapped sample ranges can be crossfaded in order that a gradual shift in timbre takes place over the region of takeover between one range and the next

in the sound at the boundary between two note ranges might be undesirable, allowing the takeover from one sample to another to be more gradual. For example, in the organ scenario introduced above, the timbre could change noticeably when playing musical passages which crossed between two note ranges because replay would switch from the upper limit of transposition of one sample to the lower limit of the next (or vice versa). In this case, a positional crossfade mode is selected by the user for the replay program, and the ranges for the different samples are made to overlap (as illustrated in Figure 3.23).

Over the range where two samples appear to be assigned to the same note number, the system mixes a proportion of the two samples together to form the output. The exact proportion depends on the range of overlap, and the note's position within this range. Normally the crossfade will result in a gradual reduction in the proportion of the lower-pitched sample and an increase in that of the higher-pitched sample as the player plays a chromatic scale upwards through the crossfade range. Very accurate tuning of the original samples is needed in order to avoid beats when using positional crossfades. Clearly this approach would be of less value when each note was assigned to a completely different sound, as in the drum kit example.

3.12.3 Velocity crossfades

Crossfades based on note velocity allow two or more samples to be assigned to one note or range of notes. This requires at least a 'loud sample' and a 'soft sample' to be stored for each original sound, and some systems may accommodate four or more to be assigned over the velocity range. The terminology may vary, but the principle is that a velocity value may be set at which the replay switches from one stored sample to another, as many instruments sound quite different when they are loud to when they are soft (it is more than just the volume that changes: it is the timbre also). If a simple switching point is set, then the change from one sample to the other will be abrupt as the velocity crosses either side of the relevant value. This can be illustrated by storing two completely different sounds as the loud and soft samples, in which case the output changes from one to the other at the switching point. A more subtle effect is achieved by using velocity crossfading, in which the proportion of loud and soft samples varies depending on the received note velocity value. At low

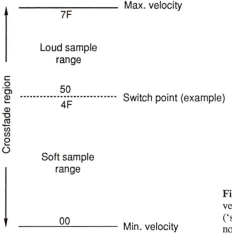

Figure 3.24 Illustration of velocity switch and velocity crossfade between two stored samples ('soft' and 'loud') over the range of MIDI note velocity values

velocity values the proportion of the soft sample in the output would be greatest, and at high values the output content would be almost entirely made up of the loud sample (see Figure 3.24).

3.13 Miscellaneous MIDI functions

Samplers are capable of responding to most of the MIDI control messages in a similar fashion to the synthesisers described in the first part of this chapter. A sampler will, for example, normally accept and act on pitch bend data, and also on modulation wheel data. Controller messages may be connected to different signal processing modules within the sampler depending on the current mapping defined by the user.

Program change messages will normally result literally in a 'program change' within the sampler: in other words, a change from one set of sample assignments and treatments to another. For example, one program number might correspond to a replay program in which drums and a bass instrument were assigned to particular note ranges, while another might assign the whole note range to a drum kit. This depends entirely on the samples in memory at any one time, and the flexibility of the individual system. At any time, a different set of samples and programs could be loaded into memory from a disk or by sampling new sounds. In a more simple sampler without the facility for multiple programs and voices, the program change could result in a change from one sound to another: rather like a voice change in a synthesiser. More recently, samplers with disk storage have been provided with the facility for loading a file from disk on the receipt of a particular program change message, after which a second message is used to define the relevant program from within the file. System exclusive implementations may also exist to control certain of the sampler's system functions, and these are normally described in individual manuals.

3.14 Sample editing

3.14.1 Overview

MIDI provides the facility, using system exclusive messages, for the dumping of sample data to and from devices in a system over the bus. The primary usage for sample dumps is in the transference of samples to a desktop computer for editing, using some of the software packages that are available for this purpose. A sample may be displayed in graphical form on the screen of the editing computer, and may be modified in various ways to change or improve the sound. Subsequently the data may be transferred back to the sampler in its modified form.

Dumping sample data over MIDI is painfully slow for all but the shortest sounds. One second's worth of 12 bit audio at a sampling rate of 30 kHz would result in 360 000 bits of data. The baud rate of MIDI is 31 250 bits per second, thus the limitation in speed can be clearly seen. Taking into account the need to format the audio sample data in a manner suitable for MIDI transmission, this one second sample would take nearly 20 seconds to transmit.

An emerging option for speeding up sample transfer now exists in the form of 'SMDI', which is a means of using the fast SCSI interface for transferring MIDI-style data. SCSI interfaces are found on an increasing number of samplers as a means of connecting hard disks and CD-ROM drives, and they are also present on many desktop computers. It should not automatically be assumed, though, that it will possible to connect the SCSI interfaces of your sampler and desktop computer together and achieve communication. Success in communication depends on the software running at both ends. Top-end computer-based sample editing packages usually allow a number of options for the dumping of sample data to and from MIDI samplers, including MIDI, RS422 and SCSI, together with configurations for many of the most widely used commercial devices. Provided that your sampler is supported for SCSI transfer by the editing package in question all should be well.

The protocol for MIDI sample dumps has now been standardised, allowing all devices which might work with sample data to talk to each other. Older systems may not conform to this standard, and it is possible that devices may incorporate their own system exclusive standards for bulk dumps of data to other devices from the same manufacturer in addition to the universal standard (these will be headed by the manufacturer's own unique ID as opposed to the universal header). In standardising the transfer of audio data a number of hurdles have to be crossed, one of which is the fact that samplers do not all work to the same quantising resolution. Some samplers use eight bits per sample, some use twelve and others use sixteen. As already explained, MIDI data bytes only have seven active bits available to carry data, which means that the audio sample data must be rather oddly divided up between words.

3.14.2 MIDI sample dump format

Sample data may be transferred either using a single MIDI link from sampler to receiving device (open loop), or with links in both directions so that the receiving device can signal correct reception to the transmitter (closed loop), as shown in Figure 3.25. The closed-loop system is a more secure way of transferring data, as the receiver can request retransmission of data if any is erroneous or lost for whatever reason. In a closed loop, the MIDI OUT of the sampler would be connected

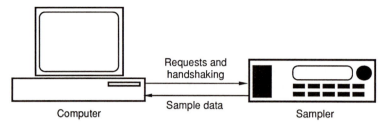

Figure 3.25 A closed-loop sample dump configuration in which a computer requests the dumping of a stored sound from the sampler

to the MIDI IN of the receiver, and the OUT of the receiver (not the THRU) would go back to the IN of the sampler.

Two types of message can be used in this protocol: principally those which carry sample data itself, and those which are for 'housekeeping' purposes (known as handshaking). Handshaking data is returned from the receiver to the transmitter in order to signal such things as whether the data has been received correctly or not. All sample dump messages are in the system exclusive format, thus they begin with &F0 and end with &F7. &F0 is followed by the universal non-realtime header, &7E, which, in conjunction with relevant sub IDs, signifies a sample dump. Most sample dump messages use only one sub-ID, and receivers should not expect to see a sub-ID #2 except in the case of more recent extensions to the sample dump standard designated by the sub-ID #1 of &05. The messages are similar to the MIDI file transfer messages (see Chapter 5) which follow a similar format.

Assuming a closed loop, the sequence of events for a sample dump would follow the following protocol. Firstly, the receiver (let us assume a desktop computer) should ask for the sample it wants. This message may also be initiated from a front panel control on the transmitter in the case of an open loop, rather than over MIDI. The message is 'dump request', and takes the form:

&[F0] [7E] [dev. ID] [03] [ss] [ss] [F7]

The sub-ID #1 of &03 signifies a dump requests, and [ss ss] is the number of the requested sample as stored in the sampler itself (LSbyte first). Thus up to 2^{14} samples could be selected by the two [ss] bytes.

The receiver (the sampler) should then check that the number of the requested sample is within the range of its stored samples, and ignore the message if it is not. If the sample number is legal, the transmission should begin, preceded by the dump header (sub-ID #1 = &01). This describes all the 'vital statistics' of the sample to be transmitted, so that the receiving device knows exactly what form it takes, how many bits there are per sample, what the sampling rate is, and so on. The header takes the form:

&[F0] [7E] [dev. ID] [01] [ss] [ss] [ee] [ff] [ff] [ff] [gg] [gg] [gg] [hh] [hh] [hh] [ii] [ii] [ii] [jj] [F7]

where [ss ss] = stored sample number (LSbyte first); [ee] = number of bits per sample (from 8 to 28); [ff ff ff] = sample period in nanoseconds (expressed as 1/sample rate; LSbyte first); [gg gg gg] = sample length (number of words of 8 to 28 bits); [hh hh hh] = sustain loop start (in number of words from sample start; LSbyte first); [ii ii ii] =

sustain loop end (as above); [jj] = loop type (00 = forward, 01 = forward/backward).

Within 2 seconds, the receiver should send one of two handshaking messages back to the transmitter: either ACK (acknowledge) if it is capable of accepting the data described by the header, or CANCEL if it is not. ACK takes the form:

&[F0] [7E] [dev. ID] [7F] [pp] [F7]

and CANCEL takes the form:

&[F0] [7E] [dev. ID] [7D] [pp] [F7]

where [pp] is the packet number (see below). In the case of this first message to acknowledge the header, the packet number is unnecessary, or could be set to zero. On receiving ACK, the transmitter should start sending packets of sample data. CANCEL should abort the exchange. If nothing is received by the transmitter within the 2 second period, it will assume that an open loop is in use and begin to send packets of data anyway.

A third acknowledgement is possible after receipt of the header: this is WAIT (&7C), which signals that the transmitter should wait until another message is received. The receiver could then send one of the other two messages when it became ready to operate. Therefore it is recommended that WAIT be sent initially, rather than nothing, if a delay in responding greater than two seconds is envisaged.

Sample data is transmitted in 120 byte packets, regardless of the size of the sample or the number of bits per sample. A packet takes the form:

&[F0] [7E] [dev. ID] [02] [kk] <120 bytes of sample data> [LL] [F7]

where [kk] is a running count of packets from 0 to 127, starting again from zero if necessary. [LL] is a checksum, which is the exclusive or function of the bytes from [7E] up to but not including the checksum byte itself. The 120 bytes of MIDI sample data represent stored sample words in 2, 3, or 4 byte groups as follows (remember that each byte can only use seven bits, because MSB = 0). For samplers using between 8 and 14 bits per sample, 2 bytes represent a sample word (60 words per packet); for samplers using between 15 and 21 bits per sample, 3 bytes represent a sample word (40 words per packet); and for samplers using between 22 and 28 bits per sample, 4 bytes represent a sample word (30 words per packet).

Audio data is transmitted MSB first within the groups of MIDI bytes, 'left justified' if the sample data does not completely fill the space available, with unused bits set to zero. In other words the MSB of the sample word should always be the first active bit of the first packet byte transmitted. Checksums of each packet are to be verified by the receiver, and the return link used to send either ACK if it is correct (see above) or NAK (not acknowledge) if it is not. NAK takes the form:

&[F0] [7E] [dev. ID] [7E] [pp] [F7]

where [pp] is the packet number. NAK should result in a retransmission of the offending packet if possible.

Not all devices may have the ability to receive packets out of sequence, although the presence of individual packet numbers makes it *possible* to transmit any single packet to a device capable of handling this. There is possible room for confusion if the number of packets exceeds 128, as the sequence of packet numbers will begin again from zero.

Although the universality of the sample dump format makes for compatibility, it may also be slower than an individual manufacturer's dump format, due to the need to accommodate and signal a wide variety of sample types.

The sample dump standard has recently been extended to incorporate the signalling of multiple loop points, and the reader is referred to the latest revision of the standard for relevant data.

3.14.3 Sample editing software

Once samples have been transferred into computer RAM they may be modified in an enormous variety of ways. Various looping possibilities have already been introduced, and it is also possible to perform such operations as equalisation, level control, dynamics processing, crossfading, mixing, time compression/expansion and pitch shifting. Some packages also allow sample data to be displayed in both time domain (waveform) and frequency domain (spectrum) forms, with the possibility for performing a Fast Fourier Transform (FFT) on the stored data so as to produce a plot of the way in which the sample's spectrum changes with time (waterfall plot).

Sample data can also be stored on the computer's disk drive in the form of a sound file, and a number of file formats are in use for this purpose, depending on the type of computer involved. Digidesign's Sound Designer formats are popular because of the wide-ranging use of Digidesign audio cards in music computers, and Apple's AIFF is also common. If the computer has the appropriate audio card built in, it will be possible to audition sampled sounds immediately they have been modified without needing to transfer them back to the sampler. It is also possible for a sample editing package to use sections of files made by a digital audio recording system, provided that they are in a compatible format, eliminating even the need to record the sounds with the MIDI sampler. The edited sample data could then be transferred to the sampler for replay only, under MIDI control.

If the editing software offers sample rate conversion, it will be possible to take a sound file made by one device at one rate and convert it to another rate for transfer to an alternative device (for cases in which devices operate at different sampling rates). The process of sample rate conversion can be performed with minimal effect on the sound quality of the audio, but there may be minor side effects in terms of noise and distortion depending on the quality of the algorithm. Clearly if the sample rate is reduced a lot there will be a noticeable reduction in audio bandwidth.

3.15 An overview of other MIDI musical devices

Although synth and sample modules are likely to be the most common devices in the MIDI music studio, a wide variety of alternative MIDI controllers are available for people who find the keyboard an unsuitable interface. Drum machines are also mentioned here, although they are really very similar to synths and samplers.

3.15.1 Drum machines

A drum machine is basically a store of synthesised or sampled drum sounds, which may be triggered by buttons on the front panel, pressure-sensitive pads, audio signals or MIDI information. Many drum machines incorporate their own sequencers. This does not prevent them from being externally sequenced, indeed this may be a better solution, but means that the drum machine must be capable of reading timing information such as clock bytes, song pointers and start/stop/continue messages, in order that it may be combined with other devices in a larger sequenced system.

Under MIDI control, a note on message may be assigned to a particular drum such that it triggers the replay of that drum every time the message is received. Because a drum is basically a 'one-shot' device: that is one which is hit and left to ring, as opposed to a device with sustained output, the note off message can usually follow the note on message quite quickly, and will not necessarily result in the silencing of the output. Some drum machines even ignore note off messages. For this reason, the MIDI OUT information of a drum machine often has very short time intervals between note on and note off messages for each note. If it is necessary to silence a percussion sound which continues after the note off message is received it may be possible to do so by sending a second note on message for the same note but with a very low velocity value.

It should be possible to program the machine with the assignment of MIDI note numbers to drum sounds, and most machines are velocity sensitive, with the velocity controlling the volume of the drum. Some drum machines enable 'pitched' drum sounds, with the transposition depending on the note number received. It is also possible to find machines which incorporate controllers such as pitch bend and modulation wheel. The MIDI OUT of the machine will carry note on and off data corresponding to drums played on the device itself, and the machine will respond to such data arriving at its MIDI IN.

A drum machine should be able to recognise song pointers, MIDI clocks, start, stop, continue, and the song select message which can be used to select one of a possible number of stored drum sequences from memory. There may also be an advantage if the drum machine reads MIDI timecode if it is to be synchronised to a sequencer which is also locked to this form of timing data.

3.15.2 Pitch-to-MIDI convertors

Devices exist which attempt to convert the pitch of an audio signal into a MIDI note number. This would, for example, allow a singer, via a microphone, to control MIDI equipment. It is not a straightforward task to determine the fundamental pitch of an audio signal which may have a complex harmonic content, unsteady pitch and background noise. The sensing and conversion process usually involves some delay, which can be artistically annoying, and low frequency sounds generally take longer to detect than high frequencies, since more time must elapse for a certain number of cycles to be measured.

A common application of such a technique is in the use of a guitar to control MIDI equipment, whereby the fundamental pitch of a string is sensed from a pickup, and translated into a MIDI note number. Note on and off messages are sent at a predetermined audio threshold. Sometimes a pitch bend lever is supplied, because it is difficult to implement sliding effects and pitches other than those of specific MIDI note numbers without a special control.

3.15.3 MIDI guitar controllers

The MIDI data format was, without a doubt, designed mainly for keyboard instruments, although attempts have been made to open up the world of MIDI control to musicians who play guitars and wind instruments. The success of these is variable. Some electronic guitar and bass instruments are specifically designed with MIDI in mind, and use a more reliable method for deriving MIDI information from the guitar than that indicated above. Rather than deriving the note number from an audio signal, the MIDI guitar senses the positions of the fingers on the fretboard, and the plucking of each string together with its 'velocity', to make up the note on message. Commonly, not much in the way of sound is produced from the instrument itself, the sound being produced from the MIDI device attached to the guitar's MIDI OUT. This could be a sampler or synthesiser.

A comprehensive guitar will output the data corresponding to each string on a separate channel in mono mode, allowing for independent pitch bend of each channel, and true portamento. It also allows for a different sound to be controlled by each string if desired, when connected to a MIDI expander unit with multiple voices.

3.15.4 MIDI-to-CV convertors

It is possible to convert some MIDI information into analogue voltages which can be used to control older synthesisers. A MIDI-to-CV (control voltage) convertor will give out a DC voltage corresponding to the note number or controller value, based on a convention of so-many-volts-per-octave for note messages, so that an analogue synthesiser without MIDI facilities can be included in a MIDI system.

3.15.5 Miscellaneous MIDI instruments

Instruments have been developed which allow MIDI data to be output from a special wind instrument, behaving something like a saxophone or a trumpet. They convert the combinations of keys pressed or sliders moved and the breath pressure into corresponding MIDI messages. This can then be used to control a synthesiser or other sound-producing device. Mono channel mode 4 is often used here, as with guitar controllers, to allow true legato playing, whereby if the fingering is changed without blowing again the note changes without entering a new attack envelope phase. Alternatively, more recent devices may make use of the legato switch message, or it may be possible to use the pitch bend message instead of a note change.

In order to obtain an effect as close as possible to that of playing a real wind instrument, a separate sound processor is sometimes used. Akai's EWI (electronic wind instrument), for example, requires that the EWI processor is connected after the output of the synth module or sampler which the EWI is controlling. This processor is really an envelope shaper which uses the MIDI breath controller to alter the gain of a VCA which alters the volume of the sound in order to improve the expressive qualities of the instrument.

There are also esoteric products such as MIDI vibes, a MIDI violin and a MIDI accordion, as well as a variety of alternative MIDI percussion controllers, but it is not proposed to cover these in detail here.

Chapter 4

Implementing MIDI in studio and lighting equipment

This chapter will describe how MIDI may be used for purposes other than those which are directly musical. Because of the success and low cost of MIDI as a means of remote control in the music business it is often convenient for it to be used in the control of other studio devices such as mixers and effects units. In this way, one computer may be used to handle the automation of a complete studio, provided that the limitations of MIDI control are appreciated. MIDI control has also been adopted with enthusiasm by the lighting business for similar reasons.

4.1 Mixer automation

4.1.1 Principles

In its most basic form automation is used for storing and recalling console fader positions and mute settings either dynamically (continuously updated) or statically (in the form of a snapshot). In analogue mixers this is normally carried out by fitting an indirect means of gain control into the fader's audio path, as shown in Figure 4.1. This may be provided as a standard feature but is often available as a retrofit option either from the mixer manufacturer itself or from a third party manufacturer of automation systems. The fader then carries a DC control voltage instead of audio, which is converted into a binary value by an A/D convertor (typically having between 7 and 10 bits of resolution). This value is read periodically by the automation CPU which addresses each of the faders in turn in order to 'scan' their positions. The fader positions are then stored in memory, along with information about the time of the reading (normally derived from a timecode signal), as well as being fed back to the audio control path of the fader itself so as to control the gain. Here a D/A convertor converts the fader position data back into an analogue control voltage (CV) which adjusts the gain of a voltage-controlled amplifier (VCA) through which the audio passes. Once a fader's position has been stored against a timecode value it can be used to control the gain of the channel on subsequent replay of a mix. The value may then be modified or updated on subsequent mix passes if required.

Similar techniques are used for digital mixers, or where the gain of an analogue signal path is adjusted by direct digital control. Instead of using A/D and D/A convertors, the fader position may be encoded directly into a digital value using an alternative form of encoder, and the gain of the channel controlled by using the value

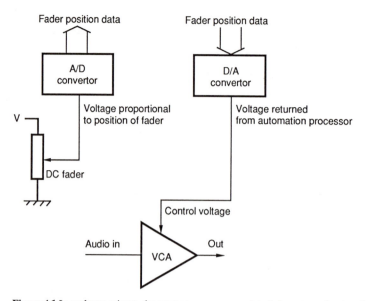

Figure 4.1 In analogue mixers, the most common approach to fader automation is to incorporate a VCA in the audio path, whose gain is controlled by a DC voltage derived from a D/A convertor. The fader's position is converted into a binary value in order that it can be stored and processed by the automation system

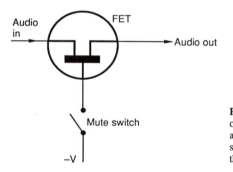

Figure 4.2 A field-effect transistor (FET) is often used as a means of muting audio in mixer automation. It has high attenuation in the 'off' state. The precise implementation depends on the type of FET and the surrounding circuit

returned from the automation CPU to adjust either a digitally controlled attenuator (DCA) in the case of the digitally controlled analogue mixer, or a digital signal processing (DSP) algorithm in the case of the all-digital mixer.

One of the advantages of such indirect control is that the value returned by the CPU to control the channel gain can be affected by other parameters. For example, one fader could be assigned as a group master and its position used to alter the gains of all the channels in the group, by combining the value read from the group master fader in an appropriate manner with the values read from the other faders in the group.

A number of means exist of displaying the gain value of the fader as returned from the CPU. VCA-based fader automation systems usually allow gain values to be displayed in a graphical form on a monitor, because the physical position of the fader often does not correspond to the actual gain of the channel. (When replaying a mix the physical fader would remain stationary but the channel gain would change depending on values returned from the stored mix.) Such systems often have nulling lights on each fader which show the direction in which the fader must be moved in order to make its position match the current channel gain. (This is necessary when entering and leaving certain update modes during subsequent passes through the mix, in order that updated fader movements take over smoothly from old data.) Alternatively moving faders may be used so that the channel's gain is indicated correctly by the fader position. In a moving fader system the gain value returned from the CPU is used to drive a motor which moves the fader to the right place. Moving faders normally need to be touch sensitive in order to allow manual control to be taken over from the motor, so as to update the fader position.

Mute switches are easy to automate because a switch is already a binary control. As with faders, the switch is separated from the means of muting and control is therefore indirect. In an analogue mixer the muting is often achieved using a switch and a field-effect transistor (FET) as shown in Figure 4.2. FETs are capable of very high degrees of attenuation in the 'off' position, and thus are good for providing effective mutes. In digital mixers the mute data controls a signal processing function designed to turn off a particular channel. The automation CPU will scan the positions of mute switches regularly and may store groups of mutes as bits of a binary word, since a switch only needs a single bit to represent its state.

A block diagram of a generalised automation system is shown in Figure 4.3. The automation CPU uses data and address lines for communication with console controls. An area of RAM is available for temporary storage of mix data, and a means of more permanent storage such as a disk drive is normally provided. (This is not necessary in a MIDI-based system, as explained below.) A timecode reader is used to determine the current position in the mix, and to 'time-stamp' automation data when it is stored.

As well as fader and mute information it is possible for virtually any other mixer control to be automated, provided that the mixer's architecture is designed so that the control's physical position and the means of control itself are available to the mixer CPU. Most conventional analogue mixers are not like this, because it requires the installation of position detectors and either VCAs or DCAs for every control on the mixer's surface, making the mixer prohibitively large, hot and expensive. Digital mixers or digitally controlled analogue mixers, on the other hand, are normally built with total automation in mind, and are usually based on a form of assignable architecture.

An assignable architecture is one in which there is less than one physical controller per control function, requiring the physical controller to be assigned to the functions as necessary. It is based on the assumption that the number of controls actually operated simultaneously on a mixer is limited, and allows the control surface to be designed with better display facilities and ergonomics which put control within easy reach of the operator. Rotary controls may be larger and continuously rotating, allowing the display of gain to be integrated with the control, either as part of the control head or around its rim, as shown in Figure 4.4. Such mixers automatically require that the CPU has access to the control of every internal audio parameter, and thus it is a relatively straightforward matter to use this information for automation purposes.

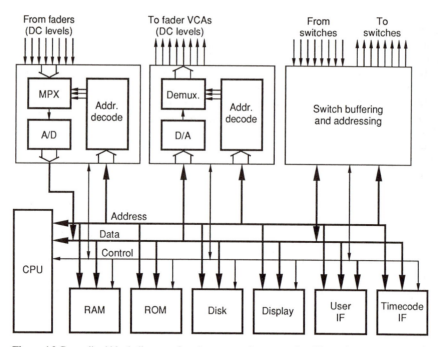

Figure 4.3 Generalised block diagram of a mixer automation system handling switches and fader positions. A DMA (direct memory access) controller may be added to speed up transfer of fader data to and from RAM. The fader interfaces incorporate a multiplexer (MPX) and demultiplexer (Demux) to allow one convertor to be shared between a number of faders. The control bus handles interrupts and enable lines to regulate data flow

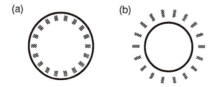

Figure 4.4 Two possible options for positional display with continuously rotating knobs in automated systems. (a) Lights around the rim of the knob itself; (b) Lights around the knob's base

4.1.2 Using MIDI in mixer automation

Before MIDI, a mixer automation system was usually an expensive dedicated product, either a console manufacturer's own, or a retrofit package from a third party designed to be added to an existing console. MIDI control has brought about the development of a number of reasonably priced automation systems for audio mixing consoles. Such automation may take the form of a MIDI interface on the mixer which carries data relating to the console's controls in the form of MIDI information, allowing that information to be handled alongside music data by a sequencer.

Alternatively it may involve the use of dedicated software to be run on a desktop computer, which perhaps displays fader levels, mute status and timecode values in a form more easy to interpret than the raw MIDI control messages displayed by a music sequencer. In the former case mixer information is treated like any other MIDI data by the sequencer and may be recorded on tracks, edited and stored. Many sequencers also have on-screen 'faders' which can be used to control the MIDI parameters of automated mixers.

There is some conjecture concerning the comparative performance of MIDI-based and conventional automation systems. This centres around the resolution available from standard MIDI controller messages and the speed of updating possible. This is discussed briefly later, but as with all such questions one must consider the relative cost-to-performance ratio. MIDI control proves adequate for mixer automation in a wide range of circumstances, and the cost of implementing it is usually quite low. Alternative dedicated systems exist and their merits should be weighed against their cost.

In order for a MIDI-based fader automation system to be possible on a conventional analogue console it is necessary for one of the following conditions to be satisfied: (a) the console already uses indirect fader gain control by means of VCAs in the signal path and a DC control voltage supplied by the fader, into which system the MIDI automation could be tapped; (b) the console may be fitted with replacement faders of a type which can be controlled by the automation system; or (c) VCAs may be housed in a stand-alone unit into which audio is patched from break points in the audio channels on the mixer, the control signal being obtained by rewiring the existing audio fader so that it becomes a DC control.

When using MIDI for mixer automation it is not necessary for the mixer end of the system to have many facilities. Indeed the mixer's automation CPU may simply be a means of scanning the faders and switches and converting the information into MIDI-compatible data. Alternatively, the mixer's CPU may handle the storage of snapshots, but allow these snapshots to be recalled under MIDI control. Digital mixers with MIDI facilities, like Yamaha's DMP-7 and DMC-1000 for example,

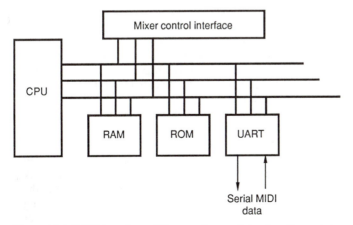

Figure 4.5 A UART is used as an I/O port to the console's automation system as a means of transmitting and receiving serial control information in the form of MIDI messages

have allowed any audio parameter to be mapped to any MIDI controller or note number, so that when a switch is pressed or a control is moved on the mixer itself the appropriate MIDI message is generated from the mixer's MIDI OUT. MIDI controller data received by such a mixer is used directly to alter the relevant mapped audio parameter. Such a mixer could be connected directly to one of the MIDI ports of a desktop computer, and its data recorded using appropriate software. All the manipulation and control of the data can take place on the computer.

Figure 4.5 shows that the automation CPU can read and write data using a serial UART for MIDI communication to and from the desktop computer. Fader movements can be represented as continuous controller information in a sequencer's edit window, as illustrated in Chapter 5. Having the mix data in a sequencer memory enables the user to edit it like any other MIDI data, with the possibility for adding and deleting mutes and applying scaling or contouring to fader levels.

4.1.3 Controller mapping

In many systems it is common for mutes to be mapped to MIDI note messages and for faders and other sliders or rotary controls to be mapped to continuous controllers. Each channel's mute control is mapped to a particular note number, and the velocity value of the note on message used to determine whether the mute is on or off. (Commonly velocities of 64_{10} and above are used for mute ons, and velocities below 64_{10} are used for mute offs.) A note on is followed soon after by a note off to limit the duration of the mute event from the sequencer's point of view. The mute function is one which must be timed very precisely, as mutes are often crucial and finely judged. The potential delay resulting in a large system should be anticipated, and if possible some provision should be made in the software for fine tuning of mute positions to take account of this.

The first 64 controller messages each carry a 7 bit data value, and theoretically these may be grouped into pairs to form 14 bit controllers. The next 32 controller messages are always 7 bit controllers. Most MIDI mixers, though, as with most synthesisers, use all these as 7 bit controllers, giving 128 steps of gain across the range of a continuously variable control such as a fader or pan control. It is possible that such coarse resolution might not be adequate, although some would say that the steps would be inaudible in terms of audio parameter change. In order to limit the audible effects of 7 bit resolution, some manufacturers introduce a ramping function into the level change, so as to smooth the transition from one gain step to another. They may also alter the fader law (the way in which gain changes with physical movement) so that the finest resolution is available at the top of the fader's travel, leaving coarser resolution at low levels.

The use of 14 bit controllers by a MIDI mixer does not necessarily mean that all 14 bits are used, since the encoders used to represent the fader position and the accuracy of gain control may only be 8 bit, but this extra bit of resolution requires the use of the whole extra controller byte and thus limits the total number of faders to 32. 14 bit controllers also make it rather difficult to edit the fader level information on an ordinary sequencer, because it is rarely possible to display combined 14 bit controllers in editing windows (they are displayed as two 7 bit controllers). For this some dedicated MIDI automation software would be needed.

An alternative approach to the problem of encoding mixer data within the MIDI format is to use system exclusive messages. These are longer than controller messages and thus less of them may be transmitted per second, but into them may be encoded any type of mixer control data split into 7 bit words. One interesting technique is to start a SysEx message and then to send control information intermittently for quite a long time, relating to a number of mixer parameters at whatever the required resolution before sending the &F7 EOX message. This avoids the need to send SysEx 'headers and footers' for every control message transmitted, thus using the available bandwidth more efficiently. It rather demands the use of separate MIDI automation software, though, since the average sequencer, although it would store the SysEx data, would make little sense of it from a display or editing point of view. Some manufacturers even recommend the use of a separate computer for MIDI mix data, in order that the MIDI input can be dedicated to mix information and not be interrupted by music data, but a fast multiport MIDI interface together with a MIDI data router like Apple's MIDI Manager should be a suitable alternative, provided that the software is compatible and the computer is reasonably fast.

Advanced MIDI controlled mixers may operate in a similar fashion to a multi-timbral synthesiser, in being able to receive on more than one channel at a time. This allows the number of controls that may be accessed over MIDI to be increased many times over. The mapping of controls to MIDI controllers may be split into banks, and all 64 continuous controller messages may be used on each MIDI channel. Alternatively, as in the Yamaha DMC-1000, one particular controller message may be used to switch between controller banks and all the following controller messages on that channel are then assumed to relate to the new bank.

The rate at which MIDI controller messages are transmitted may be user selectable. Too slow and the effect will be one of jerky step changes, but too fast and an unnecessary amount of data will be generated, thereby clogging up the bus and eating up sequencer memory. The mixer software may be able to determine automatically what is appropriate depending on the size and speed of control movement.

4.1.4 Controller chasing

When storing MIDI mix data using a sequencer it is important that the sequencer is capable of 'chasing' controller information when locating to a new point in a song (see section 5.6). This is necessary because mutes and levels may change regularly throughout a song and could be set incorrectly if one located to a new point without checking what had happened to mutes and other controllers in between.

It is less common for sequencers to chase note and other information, though, making it preferable to assign all the MIDI functions of a mixer to controller messages if possible.

4.1.5 Program changes and scene memories

MIDI program change messages are typically used to switch between snapshot or 'scene' memories in MIDI controlled mixers. Such memories may be used within the mixer to store anything from basic mute configurations to complete console setups. Program change numbers are mapped to snapshot memories, rather as controllers are mapped. Some mixers allow the user to program a crossfade time between two scene memories in order that the program change message results in

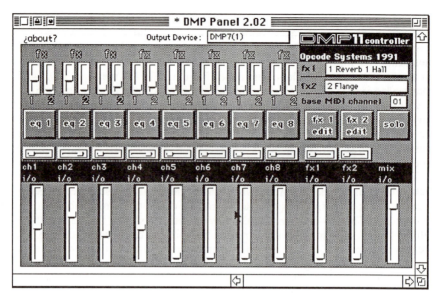

Figure 4.6 Example of a mixer setup editing display for the Yamaha DMP-11 mixer (from Opcode MAX)

a gradual fade from one lot of settings to another. This type of control is very useful in live situations, and is used widely in theatres and other such environments for the recalling of specific preprogrammed settings.

Program change messages may be generated automatically by the mixer when the particular snapshot is recalled using a front panel control, and may be stored in a sequencer in the same way as other data. Alternatively, the changes may be inserted directly into the sequence as events, and used to control the mixer on replay.

4.1.6 SysEx dumps

MIDI-controlled mixers often have a comprehensive SysEx implementation for the bulk dumping of console setup parameters. For example, it may be possible to dump the maps of controller and program changes, and it may also be possible to dump other data such as complete console setups. Using such dumps it is possible to communicate between the mixer and dedicated software which handles the display and editing of the mixer setup. An example of such a display is shown in Figure 4.6. The ability to perform this type of setup operation using a computer software package also offers the opportunity for a user to prepare the mixer for a session without needing to be anywhere near the mixer. This would allow sessions to be changed over very quickly in operation, between two entirely separate jobs, thereby minimising 'down-time' in the studio.

4.2 Effects units

MIDI control is now almost universal in outboard effects units. These units process audio so as to add reverberation and a multitude of other treatments.

4.2.1 Program changes

It is most common for a MIDI-controlled effects device to act on program change messages. These are those messages which would normally result in a voice change on a synthesiser, and which a synthesiser would usually issue when a voice is changed on its front panel. In the case of the effects unit, the received program change message may be used to change from one stored effect to another, and the device may allow the user to map program change numbers to specific effects (there are 128 program change messages). For example, a received program change message with its data byte set to &03 might be mapped to effect number 27, and would cause this effect to be selected unless reprogrammed.

This provides a powerful means of ensuring, in live performance for example, that a particular voice on an instrument is always accompanied by a particular effect, because as the voices are changed so the effects change accordingly. In a larger recording system where effects devices might be many and various, each handling a different part of a mix, a sequencer might take charge of storing and sending the program change messages to the effects at appropriate times.

4.2.2 Continuous controllers

MIDI controller messages have been used in such applications as the setting of limiter and compressor thresholds, and in controlling the gains of amplifiers within a device, as output volume controls, gain make-up controls, and so forth. It is possible again to map particular MIDI controller numbers to particular control parameters, such that the physical control device on the transmitting instrument may be selected as a wheel, pedal, slider or other analogue interface. It is then possible, when playing a keyboard, to use a redundant control device to change one or more of the effect parameters, without touching the effects unit itself.

There is also a collection of MIDI controller messages dedicated to general effects control for real-time alteration of parameters, although the default definitions of these seem more suitable for controlling the internal effects of synthesisers. These controllers are described in section 3.7.2, but the default names are open to change by individual manufacturers provided that they state the application, and could be applied to parameters such as reverberation time, early reflection time, and so on.

4.2.3 Note messages

Note on messages may be used to control the amount of pitch-shift in a 'harmoniser' or pitch-shifter. In this way, the output harmony note can be altered by playing a MIDI keyboard. It may also be possible to 'train' the effects device with musical scales from a MIDI keyboard so that it can add four-part harmony automatically.

The note on velocity value is sometimes used also to control effects parameters. In a MIDI-controlled compressor the velocity can be used in place of the audio signal amplitude to control the amount of gain reduction applied to the audio signal, thus the harder a key is hit, the 'louder' the audio is assumed to be by the compressor. This opens the door to the possibility for unusual effects which may be created by controlling the compressor parameters with a MIDI message from a device other than that which produced the audio being treated. Another example might be considered in which note off (key release) velocity could be used to control the length or level of reverberant decay in a multi-effects unit.

4.2.4 Timing and delay

It is often necessary to program an audio delay or echo which relates to the tempo of the music, so that delayed repetitions occur in time. Some effects units can pick this information up from a tap input on a key, or alternatively they can sometimes derive it from the tempo information supplied by the MIDI clock.

4.2.4 System exclusive

Effects units are not always provided with MIDI OUT sockets, as there is not very much that they would normally send to another MIDI device. It is possible, though, that a MIDI OUT could be used to dump setup parameters for user-programmed effects using a system exclusive message containing all the data from the internal registers. These could then be stored in a computer and thence onto a disk for library purposes.

4.3 MIDI data processors

It is possible to create interesting audio effects by processing MIDI control data, as opposed to audio signals. Various MIDI effects processors exist, which in their simplest application can be inserted between a device's MIDI OUT and its MIDI IN. An echo effect may be created by feeding back delayed note messages to a device's MIDI IN with ever decreasing velocity value, for example. An arpeggio effect may be created by feeding back delayed note messages which have been transposed by certain amounts (simply by shifting the MIDI note number), and if the delayed notes were repeated on a different MIDI channel then the echo could sound on a different instrument or voice. It will be left to the reader's imagination to consider what other operations might be performed on MIDI data to create musical effects.

This type of processing may also be possible as part of the MIDI data management in sequencer operating systems, but stand-alone processors may be more suitable in live applications.

4.4 MIDI-controlled audio routing matrixes

In large systems the number of audio sources and destinations will become considerable, and the routing of audio signals may become highly complicated, especially if the user envisages that regular changes to the routing may be made. Audio 'cross-point' routers exist which act as a matrix of inputs and outputs (24 in and 24 out in one example), allowing any input to be routed to any output. Combinations of cross-point selections may be stored in memory and recalled either manually or under MIDI control. Furthermore, a computer program may be written to program the cross-point selections from a remote controller, via the MIDI interface.

4.5 MIDI Show Control

For some time now, MIDI has been used in the control of lighting equipment, using continuous controller information to operate dimmers, and program change messages to recall memories containing preprogrammed 'scene' setups. The MIDI Show Control (MSC) protocol was designed as an outgrowth of this work, with a wider aim to control all sorts of live performance and entertainment equipment including lighting, audio-visual systems and theatre automation. There are even categories for such things as smoke control and explosions, and various dire warnings are given in the recommendations about the need to maintain safe practices even when using MSC!

It is based on the command structures used in existing equipment, although it is not the same as them. Rather like MIDI Machine Control (MMC) it is designed to be implemented straightforwardly on devices which already work with similar protocols and methods of operation. It could be used for applications as simple as small theatre lighting or as complicated as a multimedia show for a theme park.

4.5.1 Basic command protocol

MSC commands fall into the universal realtime category of system exclusive messages, and take the single sub-ID #1 of &02. The general format of a message is:

&[F0] [7F] [dev. ID] [02] [command format] [command] [data] ... ... [F7]

where [command format] defines the type of equipment to which the message is addressed, and [command] is the command identifier itself. A variable number of data bytes may follow the command, provided that the maximum number in a message does not exceed 128, but some messages have none at all. Device IDs &00–6F are used to address individual devices, whereas &70–7E are intended for optional groups of devices. The usual &7F ID is used to address all devices in the system. Both [command format] and [command] bytes reserve the &00 value to allow for the possibility of extensions to the protocol in the future.

A wide range of different types of equipment can be defined in the [command format] part of the message, the general categories of which are shown in Table 4.1. (For full details readers should refer to the most recent recommendations.) Commands themselves depend on individual manufacturers although there are a number of basic commands which most MSC devices would be expected to recognise, as shown in Table 4.2. Even so, the standard does not insist that a device implements any particular collection of commands. The data accompanying commands often represent cue numbers, lists or paths, and these are transmitted in a standard format which separates the different elements using a delimiter character of &00. For example, following [command]:

&[Cue number] [00] [Cue list] [00] [Cue path] [F7]

Table 4.1 General categories of MSC commands

Command format	Value (hex)
Lighting	0n (not 00)
Sound	1n
Machinery	2n
Video	3n
Projection	4n
Process control	5n
Pyro	6n

Table 4.2 Recommended basic set of MSC commands

Command	Value (hex)	Data bytes to follow
Go	01	Variable
Stop	02	Variable
Resume	03	Variable
Timed go	04	Variable
Load	05	Variable
Set	06	4 or 9
Fire	07	1
All off	08	0
Restore	09	0
Reset	0A	0
Go off	0B	Variable

although it is possible just to send a cue number, or just a number and list without a path. The numbers of cues, lists and paths may have decimal points in them (e.g.: 15.3.1) and thus a means has been included of incorporating the point by inserting the ASCII decimal point character &2E between the parts of the number. The parts of the number are represented as ASCII numbers from 0 to 9, in the format &3n where n is the number. Thus the example above of 15.3.1 as a cue number only would be represented as:

&[31] [35] [2E] [33] [2E] [31]

If the data in the command message represent a timecode value instead of a cue, the format is exactly the same as that used in the MMC and MTC cueing standards, as described in Chapter 6.

Chapter 5

Computer control of MIDI systems

This chapter deals with the heart of most MIDI systems: the computer and the software which runs on it. A desktop computer with a suitable MIDI interface can carry out a wide variety of tasks related to the control, storage and management of MIDI data, and there is now a plethora of software packages available to run on the different hardware platforms. Multimedia is also an important field of development, wherein video, audio and computer software are integrated. MIDI control can be used as part of a multimedia system to control external devices. There are a large number of different kinds of MIDI interface available for desktop computers, and their relative merits will be discussed, as will issues influencing the choice of computer for MIDI work. Alongside the actual software applications available for MIDI control have been developed a number of operating system extensions designed to tailor a computer's operating system to handle MIDI data with the minimum of delay, to deal with MIDI port drivers, and to manage the running of multiple MIDI applications on one computer whilst maintaining a degree of integration between them. These will be explained, with reference to examples from commercial systems.

It will be appreciated that every package uses its own slightly different terminology and has unique features. Software and hardware features change very frequently, and thus no attempt has been made to provide a comprehensive analysis of the characteristics of competing packages, but examples have been taken from commercial systems which represent either typical or interesting approaches to particular problems.

5.1 Hardware and software contrasted

A MIDI system will consist of both hardware and software. Microprocessor-controlled hardware needs software to run, and, as discussed in Chapter 1, software normally resides in RAM, ROM or on disks. Where mass production is the norm, and where a device is designed to perform a dedicated task repeatedly, it is normal for software to be factory programmed into ROM. This applies to most sound modules, keyboards, effects units and other 'off-the-shelf' equipment. The central controller of a MIDI system will often be called a sequencer, and the term is a hangover from earlier days when systems used simply to store strings of note data and replay them, having minimal editing facilities. Today, a sequencer is a highly sophisticated software package which is capable of wide-ranging manipulation of

MIDI data, and which would be almost unrecognisable when compared with earlier sequencers. It is more common for such software to be run on a personal computer (PC) than on dedicated hardware, but a number of dedicated sequencers are available which do not require a PC.

The PC is to be preferred to dedicated hardware as a central controller in most cases because it is capable of running a number of different software applications, whereas dedicated sequencers only perform the one task. Also, the display capabilities of dedicated sequencers are often limited whereas the PC can present a number of pages of information in a graphical form. A PC already has storage and expansion capabilities built in, and thus has the potential to grow as the user's needs grow. It may be networked with other equipment, and may also incorporate packages for recording digital audio alongside MIDI information, as described in section 5.9.

5.2 An overview of software for MIDI

Sequencers are probably the most ubiquitous of the available MIDI software packages. A sequencer will be capable of storing a number of 'tracks' of MIDI information, editing it and otherwise manipulating it for musical composition purposes. It is also capable of storing MIDI events for non-musical purposes such as studio automation, and may be equipped with digital audio recording capabilities in some cases. Some of the more advanced packages are available in modular form (allowing the user to buy only the functional blocks required) and in cut-down or 'entry-level' versions for the new user. Sequencer software is discussed in greater detail below.

The dividing line between sequencer and music notation software is a grey one, since there are features common to each. Music notation software is designed to allow the user control over the detailed appearance of the printed musical page, rather as desktop publishing packages work for typesetters, and such software often provides facilities for MIDI input and output. MIDI input is used for entering note pitches during setting, whilst output is used for playing the finished score in an audible form. Most major packages will read and write standard MIDI files, and can therefore exchange data with sequencers, allowing sequenced music to be exported to a notation package for fine tuning of printed appearance. It is also common for sequencer packages to offer varying degrees of music notation capability, although the scores which result are rarely as professional in appearance as those produced by dedicated notation software.

Librarian and editor software is used for managing large amounts of voice data for MIDI-controlled instruments. Such packages communicate with MIDI instruments using system exclusive messages in order to exchange parameters relating to voice programs. The software may then allow these voice programs or 'patches' to be modified using an editor, offering a rather better graphical interface than those usually found on the front panels of most sound modules. Banks of patches may be stored on disk by the librarian, in order that libraries of sounds can be managed easily, and this is often cheaper than storing patches in the various 'memory cards' offered by synth manufacturers. Banks of patch information may be accessed by sequencer software in order that the operator may choose voices for particular tracks by name, rather than by program change numbers. Sample editors are also available, offering similar facilities, although sample dumps using system exclusive are not really recommended, unless they are short, since the time taken can be excessive.

Sample data can be transferred to a computer using a faster interface than MIDI (such as SCSI) and the sample waveforms can be edited graphically.

Amongst other miscellaneous software packages available for the computer are MIDI mixer automation systems, guitar sequencers, MIDI file players, multimedia authoring systems and alternative user interfaces. Development software is also available for MIDI programmers, as described in section 5.10, providing a programming environment for the writing of new MIDI software applications. There are also a number of software applications designed principally for research purposes or for experimental music composition.

Since this book is principally concerned with explaining the MIDI standard and its implementation, only limited coverage will be given to software applications and their features.

5.3 Choosing a computer for MIDI work

Anyone intending to do serious work in the MIDI environment will need to consider what is the most suitable computer for their purposes. Currently there are three main platforms used for music purposes – the Apple Macintosh, the PC and the Atari ST or Falcon – although there are a number of other machines such as NeXT, Amiga and Archimedes used less frequently. Increasingly, though, the hardware platform is becoming less important because the major MIDI software companies are making their products available on more than one of the major platforms. Some implementations, though, may be better than others. The forthcoming 'Power PC' machines will be able to emulate many of the most widely used operating systems, possibly making the operating system issue somewhat less relevant. It has to be said, though, that to date the largest amount of music software has been developed for the Macintosh and Atari platforms, and thus it is perhaps to be expected that this large installed base will be one of the major factors determining future development. These computers have benefited from a 'user-friendly' graphical interface – a feature which appeared rather later in the form of 'Windows' software on the PC – which is ideal for MIDI applications. A number of the examples given in this chapter are taken from the Macintosh, but similar principles apply on other platforms.

Clearly the speed of the computer is an important determining factor in any prospective purchase, and the maxim should always be 'buy the fastest machine you can afford', but the question often arises as to 'how fast is fast enough?'. If you only need fairly simple sequencing, then virtually any of the current 'base level' machines will do, whereas if you intend to run multiple packages with many running at the same time a larger, more powerful machine will be needed. The things that are visibly most affected by speed are operations such as screen redrawing and response time to commands, especially if many colours are displayed. A faster machine may also be able to handle real-time MIDI control with greater ease, since it will have more processor time to spend on each application, thus reducing any latency in the handling of control data.

Where digital audio is to be recorded alongside MIDI information it is important to have a fast computer, especially when a package uses some of the computer's processing for writing audio data to disk and other audio-related tasks such as calculating crossfades. The displays of waveform generated by audio editing packages take a lot of calculation during redrawing, and this applies also to music notation packages which are notorious for requiring large amounts of processor time during screen redrawing.

There is a tendency in the computer world for software applications always to expand their capabilities as hardware speed increases, and thus a machine which was fast last year may appear slow next year when attempting to run the latest software. For this reason it is really impossible to say such things as 'a Macintosh Quadra 950 with 20 Mbytes of RAM is the most suitable computer for this or that application', since although this may be true today it will almost certainly not be tomorrow.

Atari has been very successful in the MIDI field by making a computer with a built-in MIDI interface, and it has been surprising to see what a powerful selling feature this was. Nonetheless it is a relatively simple matter to add an interface to any other desktop computer for a moderate sum of money, and in any case many of today's applications would be restricted by the single built-in port of the original Atari, demanding the flexibility offered by multiport MIDI interfaces (discussed in section 5.4.2).

5.4 Interfacing a computer to a MIDI system

In order to use a computer as a central controller for a MIDI system it must have at least one MIDI interface, consisting of at least an IN and an OUT port. (THRU is not strictly necessary in most cases.) Unless the computer has a built-in interface, as found on Atari machines, some form of third-party hardware interface must be added, and there are many ranging from simple single ports to complex multiple port products. The following discussion aims to highlight the differences between different hardware interfacing approaches, describing their relative merits.

5.4.1 Single port interfaces

A typical single port MIDI interface may be purchased for between £50 and £100 at the time of writing, and will be connected either to one of the spare I/O ports of the computer, or plugged into an expansion slot. On the Macintosh, for example, a MIDI interface is normally connected directly to one of the two serial ports (which are multi-purpose RS-422 type interfaces capable of operating at rates up to a few hundred kilobits per second), as shown in Figure 5.1(a). Any software then asks the user to configure the serial ports to suit the interface, requiring him to choose which of the two serial ports will be used and what the notional master clock speed of the UART will be (Figure 5.1(b)). (The choice of clock speeds is something of a hang-over from earlier days, when a 1 or 2 MHz clock speed was the master clock in contemporary computers.) Serial data arriving at the interface's MIDI IN is then converted into a pseudo RS-422 electrical format and fed to the computer's serial port to be treated like any other external serial data. Outgoing serial data is converted into the MIDI current loop electrical form.

MIDI interfaces for the PC have taken a somewhat different development line. The built-in DOS and BIOS routines for the PC limited serial communications to 9600 baud, requiring MIDI software writers to bypass these routines and write directly to the serial ports. The hardware specification for PC serial ports should allow them to be operated at a high enough rate for MIDI, but this does not always turn out to be the case in practice. For these reasons, and because of the relatively coarse timing facilities available in the PC operating system, the Japanese company Roland developed an interface for the PC called the MPU-401 which itself

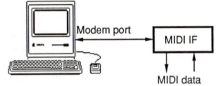
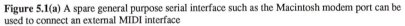

Figure 5.1(a) A spare general purpose serial interface such as the Macintosh modem port can be used to connect an external MIDI interface

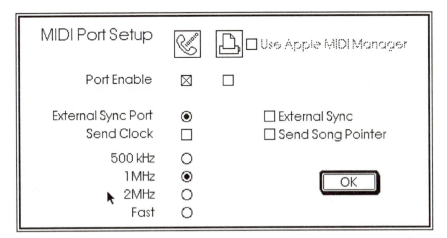

Figure 5.1(b) The interface is configured in software, using a utility such as that shown here

contained a microprocessor whose job was to handle many of the MIDI processing tasks concerned with sending and receiving note data and synchronisation information. The interface ICs were licenced to other developers, and consequently the MPU-401-compatible interface has become something of a standard for the PC. Most PC software will be able to address it, although it is truly only a single port interface even when it sports more than one MIDI OUT (all the ports carry the same information).

MPU-401 compatible interfaces connect directly to a selection of lines on the data, control and address buses of the PC, and use CPU interrupts to transfer MIDI data to and from the PC's memory. The MPU interface occupies either a memory or I/O mapped port address which may be addressed as port 0 (or anything from 0 to 3 depending on the number of interfaces). The processor on the interface acts as a 'background processor', allowing the computer's main CPU to handle its normal tasks of screen update, user interface communications and disk storage without undue delays, whilst storing or playing MIDI data. Such an interface will also handle simple synchronisation using the older timing signals of drum click and FSK (see Chapter 6). The original MPU-401 is now no longer available from Roland, and has been replaced by the MPU-IPC. There is also an IBM PS/2 expansion card available. A number of other manufacturers also make MPU-401 compatible interfaces.

Modern PCs have serial interfaces that will operate at a high enough data rate for MIDI, but are not normally able to operate at precisely the 31.25 kbaud required, thus rendering them virtually unusable for direct translation into the MIDI electrical format. Nonetheless, there are a few external interfaces available which connect to the PC's serial port and transpose a higher serial data rate (often 38.4 kbaud) down to the MIDI rate using intermediate buffering and flow control (see Figure 5.2). These would be addressed differently by MIDI software, compared with the MPU-401, requiring MIDI data to be routed directly to the serial port rather than writing it to an internal expansion port.

An alternative approach with the PC is to install a non-MPU-compatible card which drives a MIDI interface. This would require suitable software, capable of addressing the specific interface.

Therefore, to summarise, the main approaches to single port MIDI interfacing are: (a) to connect directly to a serial port running at the correct MIDI baud rate, converting the data to and from the MIDI electrical format; (b) to install an expansion card into the computer which handles MIDI communications; or (c) to connect directly to a serial port running at a higher rate than MIDI and to divide it down to the lower rate, converting data to and from the MIDI electrical format. It is vital to ensure that your chosen MIDI interface actually works with the software you intend to run – particularly on the PC where there are so many options. (See section 5.4.5, below.)

An interesting approach to single port interfacing has been taken by Yamaha (also emulated by others), in manufacturing a sound module with a serial interface that may be connected directly to either a PC's or a Mac's serial interface. The sound module, whose back panel is shown in Figure 5.3, then also acts as a MIDI interface

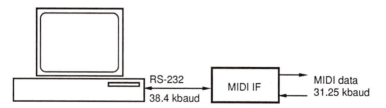

Figure 5.2 A PC may not be able to run its RS-232 serial interface at 31.25 kbaud, so the MIDI interface performs the function of rate translation to and from the higher 38.4 kbaud rate of the computer

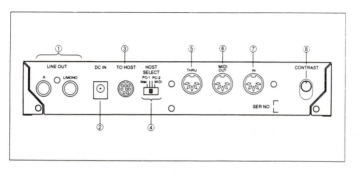

Figure 5.3 Rear panel of the Yamaha TG100, a sound generator which doubles as a computer MIDI interface, showing MIDI sockets, serial remote socket and switch for different interface types

for the computer, allowing data from the MIDI IN to be sent back to the computer, and from the computer to the OUT. A switch configures the serial port for the appropriate computer. The two PC modes are designed to deal with different serial interface data rates, since there is actually a Japanese PC clone which will operate its serial port at the MIDI data rate. The serial interface will also control the sound module directly without any MIDI connections.

5.4.2 Multiport interfaces

The MIDI protocol only allows 16 receiver channels to be addressed directly. Provided that electrical conditions were properly controlled it would be possible to connect a number of receiving devices to a single port (as discussed in Chapter 7), but the limit of 16 channels would remain. For this reason, multiport interfaces have become widely used in MIDI systems where more than 16 channels are required, and they are also useful as a means of limiting the amount of data which is sent or received through any one MIDI port. (A single port can become 'overloaded' with MIDI data if serving a large number of devices, resulting in data delays, as discussed in Chapter 7.)

An example of a simple approach to multiple port MIDI interfacing is seen on the Apple Macintosh (Mac), since the Mac has two almost identical serial ports (the printer port and the modem port). In such cases a single port MIDI interface could be connected to each of the serial ports, as shown in Figure 5.4, thereby doubling the number of channels which might be addressed. This requires software capable of addressing 16 channels via each of the independent ports, and most of the advanced packages can do this. Some require the user to decide which of the two

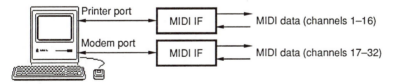

Figure 5.4 The Macintosh computer has two serial ports, each of which may be connected to an independent MIDI interface. The effect is to double the number of addressable MIDI channels provided that the software is capable of sending data to both ports

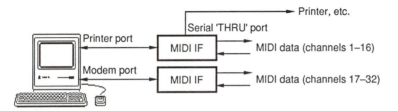

Figure 5.5 A serial 'thru' switch is sometimes provided on an external MIDI interface so that the serial data can be daisy-chained on to a non-MIDI serial device such as a printer. This saves repeated re-plugging of cables

ports is to be used for MIDI. Clearly this is only an option if the serial ports are both available. If one of them is required for other purposes then it may be possible to obtain an interface with a serial 'THRU' switch, which allows the serial data from say the printer port to be connected either to the MIDI interface or looped through to another serial port which could be connected to a serial peripheral such as a printer or modem (see Figure 5.5). Quite often, one of the Mac's serial ports is used for networking, and this renders that port unavailable for MIDI.

A more advanced approach involves the use of an external MIDI interface which has a number of independent MIDI OUT ports (each with its own UART). Such an interface is connected to the host computer using either a parallel or serial I/O port, or using an expansion card, as shown in Figure 5.6. The serial option is used on the Mac, whereas the parallel option is sometimes used on the Atari. An expansion card or parallel I/O is normally used on the PC. The principle of such approaches is that data is transferred between the computer and the multiport interface at a higher speed than the normal MIDI rate (called 'Fast Mode' on the Mac), requiring the interface's CPU to distribute the MIDI data between the output ports as appropriate, and transmit it at the normal MIDI rate. This requires a certain amount of 'intelligence' in the interface, and the routing of data is normally performed under the control of the host computer which will configure the MIDI interface according to the studio setup designed by the user.

Some older intelligent routing matrixes had multiple ports, but only had a single MIDI input to connect them to a computer's MIDI OUT. These matrixes or 'patchers' performed various filtering, processing and routing functions which ensured that only certain MIDI data was carried over certain cables, but these were not true multiport computer interfaces in the modern sense and did not result in an expansion of the number of channels available – they simply helped in the wiring and organisation of larger systems. High speed communication between the computer and a multiport interface eliminates the data 'bottleneck' which would probably occur if this connection were only made at the normal MIDI data rate, and allows data for more than 16 channels to be transferred. Most multiport systems allow up to 16 channels to be addressed by each of the MIDI OUTs, thus expanding the total number of channels to 16 times the number of ports. In such systems it is common for each instrument to be connected to its own port, both IN and OUT, as described in Chapter 7, which is especially useful with multi-timbral sound modules capable of operating on all 16 channels simultaneously. It also allows any instrument to be used as a controller, and makes it possible for instruments to send system exclusive information back to the computer. Multiport interfaces will often allow data received from more than one port to be merged, or allow recording from more than one source at a time.

An interesting commercial example of a multiport MIDI interface exists in the form of Opcode's Studio 5, pictured in Figure 5.7. It has two serial ports which transfer data to and from the Mac, as shown in Figure 5.8, and these can be 'THRU'd' if required to connect other serial devices (although the Studio 5 then loses this port for MIDI communications). Serial communication between the Studio 5 and the computer normally operates at a higher rate than MIDI, and the two serial ports provide a greater data throughput capacity than just one. The control software shares the data between the two serial ports in order to optimise flow, and it is therefore not possible to say that one serial port serves only certain MIDI ports. There are 15 MIDI ports with two on the front panel for easy connection of 'guest' devices or controllers that are not installed at the back, and the software which controls the Studio 5 allows the user to configure

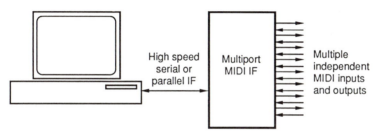

Figure 5.6 A multiport MIDI interface is connected to the computer using either a serial or parallel interface running a number of times faster than the MIDI data rate

Figure 5.7 Front panel of the Opcode Studio 5 multiport MIDI interface showing two MIDI ports for 'guest' instruments, and LEDs to show transmit and receive activity on each port

Figure 5.8 Rear panel of the Opcode Studio 5, showing Macintosh serial ports and serial THRU ports, as well as MIDI interfaces, timecode interface and other external connections

the studio so that the computer 'knows' which instrument is connected to which port. Applications then address instruments rather than ports or channels. This is discussed further in section 5.6.1. The Studio 5 also incorporates other facilities for manipulating and routing MIDI data which will be discussed in Chapter 7. It also has a timecode port in order that synchronisation information can be relayed to and from the computer (see Chapter 6).

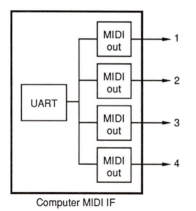

Computer MIDI IF

Figure 5.9 Multiple MIDI outs derived from a single UART normally carry the same data as each other. The approach simplifies wiring but does not expand channel capacity

5.4.3 How many OUTs?

It is not unknown for a device to appear to be offering a number of OUTs when really all the ports are carrying the same information. A single UART is connected to a number of OUT ports in parallel in this case, as shown in Figure 5.9. This can be useful for connecting a number of instruments without needing a daisy chain of THRUs between them, but does not expand the number of channels available or allow selective transmission of data over certain cables.

5.4.4 Ports versus channels

Users often confuse physical ports with MIDI channels when a multiport interface is attached to the computer. A physical port is a socket on the interface to which a MIDI device is connected, whereas the MIDI channel is only defined in the data message which is sent through that port. You could have 16 ports each of which addressed 16 channels. On each port, the MIDI channels would be called channels 1–16. When a MIDI operating system such as OMS is used, the port may be named so that the user does not have to remember which port corresponds to each instrument (assuming that each instrument has its own dedicated port).

5.4.5 Interface drivers

Quite commonly, a software driver is provided for an external MIDI interface or expansion card. This driver is used by the computer's operating system to address the MIDI interface concerned, and it takes care of handling the various routines necessary to carry data to and from the physical I/O ports. The MIDI application therefore simply talks to the driver, and it follows that you must have the correct driver installed for the interface which you intend to use. High-end MIDI software usually comes complete with a number of drivers and configuration documents for the most popular MIDI interfaces.

5.5 MIDI operating systems

In order to manage the MIDI data which is used by software applications, some manufacturers have designed operating system extensions which enhance the system capabilities of a computer. Such extensions typically operate in the background and are unobtrusive in normal operation. Features vary, but can include keeping a map of the MIDI devices connected to a multiport interface, performing filtering and processing of MIDI data for certain input and output ports, synchronising multiple MIDI applications and optimising the throughput of MIDI data to ensure optimum timing accuracy.

Two examples of such operating system extensions are Steinberg's MROS (MIDI Real-time Operating System), available for the Atari, PC and Mac, and Opcode's OMS (Opcode MIDI System), currently only available for the Mac. Apple's MIDI Manager is also an interesting example of a way in which MIDI information may be internally routed between applications without the need for external connections. These are not the only examples of such approaches in existence, but they offer useful illustrations of the concepts involved and are typical of the approaches used by other software.

5.5.1 MROS

MROS was originally designed for the Atari, and was principally intended to allow multiple MIDI applications to run simultaneously with optimal timing of events. It also handles the routing of MIDI data either to and from the built-in MIDI port of the Atari or to and from the drivers which handle external multiport interfaces such as Steinberg's MIDEX and SMP-24. In effect, MROS provides virtual MIDI ports which can be combined with powerful MIDI processing to alter the information entering or leaving the computer.

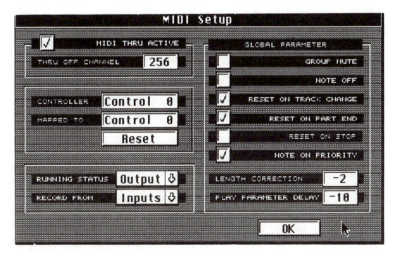

Figure 5.10 Example of MROS configuration page

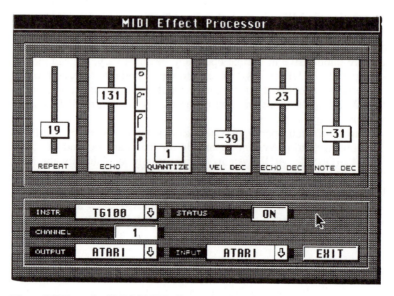

Figure 5.11 Example of MROS MIDI effects processing window

MROS is currently only used by Steinberg applications, and some example screens showing its features are shown in Figure 5.10 and 5.11.

5.5.2 OMS

OMS acts as a system integrator mainly for Opcode's software products, and it may become more widely used by other manufacturers. Opcode has decided to make OMS a more widely available tool, renaming it the 'Open MIDI System', encouraging other manufacturers to make their software OMS compatible. (It should be noted that there is currently something of a battle between Opcode and its main competitor, Mark of the Unicorn (MOTU), since MOTU is also promoting its Free MIDI System in a similar fashion.) OMS allows the user to configure the studio setup in software to represent the physical MIDI connections which exist. With a multiport interface the layout of the connections is configured in a window which might look something like the diagram in Figure 5.12. Here the user has defined instruments which are connected to each physical port, and each instrument's definition includes information about the type of data it receives and transmits, as shown in Figure 5.13. It is possible to indicate which MIDI channels an instrument will receive on, and whether or not it transmits and receives synchronisation data.

OMS also allows the definition of external 'patchers' and the creation of 'virtual instruments' and 'virtual controllers'. External patchers are MIDI routing matrixes which serve to connect one MIDI input to a number of MIDI outputs, in order to expand the number of devices which may be connected to a single MIDI port without the need for THRU chains. The patcher setup is configured in OMS as shown in Figure 5.14(a), defining the instruments connected to the patcher and

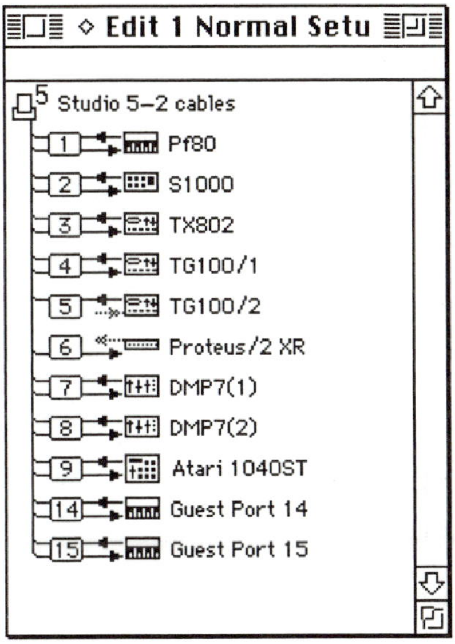

Edit 1 Normal Setu

Studio 5–2 cables

1 Pf80
2 S1000
3 TX802
4 TG100/1
5 TG100/2
6 Proteus/2 XR
7 DMP7(1)
8 DMP7(2)
9 Atari 1040ST
14 Guest Port 14
15 Guest Port 15

Figure 5.12 Example of an OMS studio setup document showing which devices are connected to each MIDI port

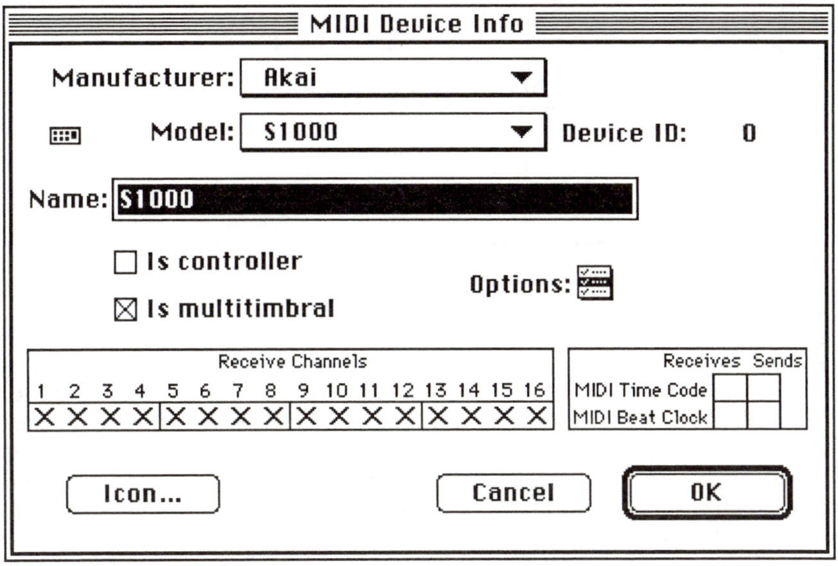

MIDI Device Info

Manufacturer: Akai

Model: S1000   Device ID: 0

Name: S1000

☐ Is controller
☒ Is multitimbral   Options:

Receive Channels																	Receives	Sends
1	2	3	4	5	6	7	8	9	10	11	12	13	14	15	16	MIDI Time Code		
X	X	X	X	X	X	X	X	X	X	X	X	X	X	X	X	MIDI Beat Clock		

[Icon...]   [Cancel]   [OK]

Figure 5.13 Each device in the OMS setup is described in such a window as this. Here the user determines what MIDI information is received and transmitted by the device, and sets the SysEx device ID

allowing certain program change messages to alter the routing configuration of the patcher. (Most external patchers will allow source to destination routings to be stored in memories, and then to be recalled on receipt of a program change number.)

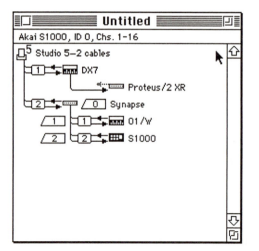

Figure 5.14(a) Example of an external MIDI patcher connected to port 2 in the OMS studio description. A MIDI THRU from the DX7 to the Proteus is also shown

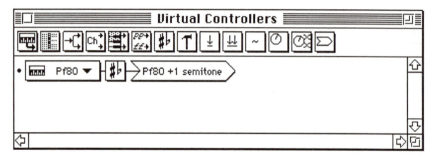

Figure 5.14(b) A simple example of a virtual controller created by transposing the note messages sent from a real instrument. The transposed instrument may then be used in its own right

Virtual controllers are the outputs of real instruments which have been processed or modified by OMS, either in order to produce a special effect, or because the instrument's MIDI features are not as required by the user. Such processing may involve such operations as filtering of data, transposing notes, mapping controllers or program changes to other numbers, splitting note ranges, producing echo loops and so on. The virtual controller may then be treated like any other OMS device. An example is shown in Figure 5.14(b). A virtual instrument is created in a similar manner.

Once a studio setup has been defined in this manner, all OMS compatible applications have access to the information. In a sequencer package, for example, the user may select an instrument for a particular track by name, and it is also possible to select which instrument(s) will act as sources for MIDI data to be recorded. It is no longer necessary to worry about which physical ports they are connected to. If the studio setup is changed then all OMS applications may be updated with the new conditions automatically.

5.5.3 Apple MIDI Manager

MIDI Manager allows multiple MIDI applications to run simultaneously on the Mac, with shared access to synchronisation data and to the computer's serial ports. Without MIDI Manager, the serial ports are normally only available to one application at a time (although OMS tackles this problem for OMS-compatible applications). MIDI Manager is configured using an application called PatchBay, an example window of which is shown in Figure 5.15. Here each open application is represented by an icon, and each icon has a number of input and output 'ports'. The small triangles are ports for normal MIDI data, and the little clocks are ports for synchronisation data (see Chapter 6). In this example, OMS is also running on the machine, and thus MIDI Manager is operating 'in collaboration' with OMS. The OMS port shown in the PatchBay example represents the software driver which talks to the external multiport MIDI interface, and is separately configured to route specific MIDI Manager ports through to the various OMS input and output devices (as configured in the OMS setup). Without OMS, this icon would be the conventional Apple MIDI Manager Driver port which fulfils a similar function.

If one wished to synchronise one application to another, a connection would be made between the clock icons of the applications concerned. Similarly, clock data could be connected from the OMS clock source in order to route timing data from the outside world to applications running on the computer. (The latter approach would be used when an external source of sync information such as a drum machine or timecode synchroniser was being used to control a computer sequencer, for example). Applications may have a number of MIDI Manager input and output 'ports', and these must be considered as 'virtual ports' since they do not exist as physical entities. They should be distinguished from the physical ports of an external MIDI interface, but they are rather like an internal 'software' version of the

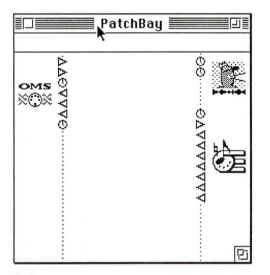

Figure 5.15 Example of Apple's MIDI Manager PatchBay application which allows MIDI data and synchronisation information to be routed between applications and MIDI interface drivers using virtual 'cables'. Each icon represents either an application or a driver

same thing. The multiple MIDI Manager ports allow selective routing of data between applications, and the applications themselves normally allow the user to select which virtual output port is to be used by which track or instrument. A similar concept is used in software MIDI routing environments used by other manufacturers and platforms.

5.5.4 Inter-application communications

It is becoming common for computer operating systems to allow active links between applications which are open at the same time. These are sometimes known as inter-application communications (IAC) or 'hot links'. Using hot links it is possible for MIDI applications to exchange information which is common to them, and this is particularly useful when two or more applications share patch libraries or need access to the same MIDI data. On the Macintosh, for example, IAC is used in some MIDI software to allow one application to synchronise directly to another without using MIDI Manager (see above).

5.6 Introduction to sequencer concepts

The section that follows is intended to provide an overview of some of the most common concepts associated with MIDI sequencers. This is an introductory explanation, showing how MIDI is implemented in sequencers, and the concepts explained here may be considered as 'core' concepts which apply almost no matter what the product or which the revision of the software. Readers who wish to find more detailed coverage of sequencer implementations and their applications are recommended to refer to Yavelow's excellent *Music and Sound Bible* as listed in Appendix 2. Examples are taken from commercial packages, to illustrate specific points, but clearly individual packages will differ in matters of terminology and detail. A lot of the examples illustrated here are taken from Opcode software, but similar concepts are found in other manufacturers' products.

5.6.1 Tracks, channels, instruments and environments

A sequencer may be presented to the user so that it emulates to some extent the controls of a multitrack tape recorder. The example shown in Figure 5.16 illustrates this point, showing the familiar transport controls in the top left hand corner. There are advantages in this approach, since it is a familiar interface for many users, but some clarification is required regarding the way in which the 'tracks' of this MIDI recorder can be made to relate to MIDI channels, as it is possible, for example, to encounter sequencers with many more tracks than there are MIDI channels.

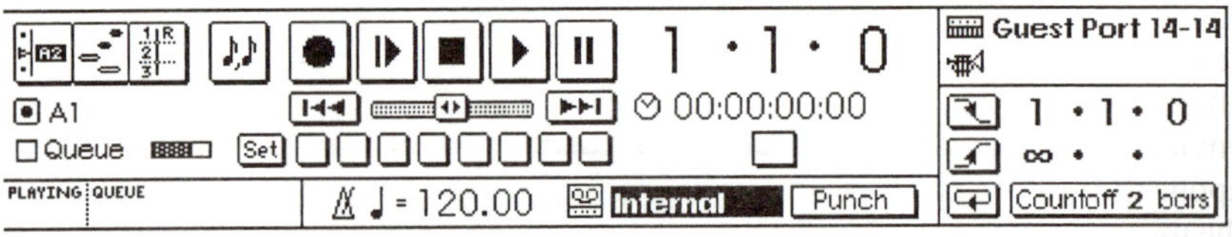

Figure 5.16 Example of transport controls in a sequencer window (Opcode Studio Vision)

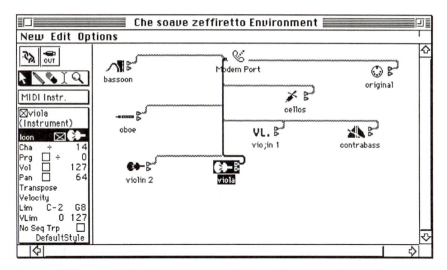

Figure 5.17 Simple example of a Notator Logic 'environment' window in which MIDI instruments, their routings and processing, may be defined

A track is simply a way of presenting memory space to the user, and it also helps in the compartmentalisation of information so that particular sections of material may be assigned to particular tracks. There is also the conceptual advantage that a composition or recording is built up by successively overlaying more and more tracks, all of which may be played together at the same time. Unlike the tape recorder, the MIDI recorder's tracks are not fixed in their time relationship and can be slipped against each other, as they simply consist of data stored in the memory. On older or less advanced sequencers, the replay of each track is assigned to a particular MIDI channel, and this results in data from that track being output with its status bytes set to that channel number. It may not matter what channel was operative when the track was recorded, as the sequencer allows the user to change it on replay, perhaps even during the song. More recent packages are considerably more flexible in this respect, offering an almost unlimited number of virtual tracks and allowing tracks to be assigned to virtual instruments. A track can contain data for more than one channel. Using a multiport MIDI interface it is possible to address a much larger number of instruments than the basic 16 channels of MIDI allowed in the past, and operating systems such as OMS allow these instruments to be chosen by name rather than by channel or port number.

In a typical modern sequencer, instruments are often defined in a separate 'environment' which defines the physical instruments, the ports to which they are connected, any additional MIDI processing to be applied, and so forth. The OMS environment setup was shown earlier, and Figure 5.17 shows an example of an alternative approach used by EMagic in Notator Logic. (The Notator Logic environment is set up solely within the application, whereas the OMS environment applies to all OMS applications currently open.) When a track is recorded, therefore, the user simply selects the instrument to be used, and the environment takes care of managing what that instrument actually *means* in terms of processing and routing.

5.6.2 Input and output filters

As MIDI information is received from the hardware interface(s) it will be stored in memory, and, unless anything is done to prevent it, all the data that arrives will be stored. If memory space is limited it may be helpful to filter out some information before it can be stored, using an input filter. This will be a sub-section of the program which watches out for the presence of certain MIDI status bytes and their associated data as they arrive, so that they can be discarded before storage. The user may be able to select input filters for such data as aftertouch, pitch bend, control changes and velocity information, among others. Clearly it is only advisable to use input filters if it is envisaged that this data will never be needed, since although filtering saves memory space the information is lost for ever. Filtering may also help to speed up the MIDI system on replay, owing to the reduced data flow, although it may slow down the computer during recording.

If memory space were not limited, it would be possible to store all the data that arrived and filter it at the output if it became necessary to prevent certain data from being transmitted on replay. Output filters are often implemented for similar groups of MIDI messages as for the input filters. It may also be possible to filter out timing and synchronisation data for those devices which do not require it. Some input and output filtering may also be performed in the MIDI operating system or environment, in which it is often possible to determine which instruments transmit and receive particular types of MIDI data, in order to limit the data flow over individual MIDI cables to a minimum. Delays in MIDI systems are discussed in greater detail in section 7.4.

5.6.3 Timing resolution

The resolution to which a sequencer can store events that are received while recording is a parameter which varies between systems, as mentioned above. Since MIDI is an asynchronous interface, data bytes may be transmitted from a master keyboard to a sequencer at any time, not limited to particular clock intervals, although the replay of a long string of data could involve some such limitations if the bus was approaching saturation point (see Chapter 7). This 'record resolution' may vary between anything from around 96 ppqn to 960 ppqn, with high-end systems typically offering the greatest resolution.

The effect of this on the timing of notes and other events depends on the current tempo, but an example could be taken using a tempo of 120 beats per minute, at which a sequencer with a resolution of 96 ppqn would only resolve to 20.8 milliseconds, whilst one with a resolution of 480 ppqn would resolve to 4.1 milliseconds. The quoted resolution of sequencers, though, tends to be somewhat academic, depending on the operational circumstances, since there are many other factors influencing the time at which MIDI messages arrive and are stored. At low tempi, stored music can exhibit the audible effects of poor sequencer timing resolution in the form of delays between the notes in a chord, for example, since the MIDI messages corresponding to the notes in that chord would have arrived serially (one after the other) during recording, and would have been stored at the closest event interval available, possibly forcing some notes in the chord to a different event time than the others. Timing resolution also influences the success of operations such as rhythm quantisation as described below. Some sequencers have editing operations which look for chords automatically, forcing notes on a given channel which occur within a certain time window to the same event time.

The record resolution of a sequencer is really nothing to do with the timing resolution available from MIDI clocks or timecode. The sequencer's timing resolution refers to the accuracy with which it time-stamps events and to which it can resolve events internally. Most sequencers attempt to interpolate or 'flywheel' between external timing bytes during replay, in an attempt to maintain a resolution in excess of the 24 ppqn implied by MIDI clocks (see Chapter 6).

5.6.4 Recording modes

A track may be recorded in a number of different ways, depending on the operational requirement. Terminology varies, but the concepts remain similar. In 'replace' mode, a new recording on a particular track will obliterate anything which was previously recorded on that track in that place, whereas an 'overdub' mode will allow new material to be added on top of old material. The point at which record mode is actually entered usually either depends on an event such as a note being pressed, or alternatively the sequencer will 'count you in' with a defined number of bars of metronome click. It is normally possible to define the range of bars to be re-recorded, and recording will begin automatically at the 'in point', dropping out at the 'out point'.

Loop recording is a function borrowed from drum sequencers, and it is particularly useful in building up a rhythmic pattern which consists of a number of different elements. Loop recording allows a predefined number of bars to be cycled over and over, with an overdub taking place on each pass. On each pass additional material may be recorded. Taking a drum loop as an example, a kick drum could be recorded on the first pass, a snare on the second, a hi-hat on the third, and so on, gradually building up the backing rhythm for a song. During loop recording it is normally possible to press a key to cancel the last recorded pass over the loop, allowing the composer to keep trying things until he is happy with the result. When the loop is satisfactory, another key can be used to 'fix' the recording. Step time recording modes allow the user to enter events in a non-realtime fashion, by pressing keys and entering durations of notes one step at a time. The pitches are often entered from a music keyboard, whilst the durations may be selected from the computer keyboard. This can be useful for programming sequences which are not possible to play in real time. It is also useful for entering musical scores in notation packages.

Depending on the hardware interface and operating system it may also be possible to record from more than one source at a time, and this can be valuable in live recording situations where more than one musician is generating MIDI information. Even with a single port interface it may be possible to merge two sources externally and treat them as one data stream. Sometimes the resulting data is recorded onto one sequencer track, but it may be possible to separate the independent streams afterwards using a special sequencer function which splits off each channel to a separate track. Multi*channel* recording refers to the ability of the sequencer to record MIDI information on more than one MIDI channel at a time, probably with all channels being stored on the same track. Multi*track* recording refers to an ability to record each channel on a separate track in real time.

5.6.5 Displaying and editing MIDI information

A sequencer is the ideal tool for manipulating MIDI information, and this may be performed in a number of ways depending on the type of interface provided to the user. The most flexible is the graphical interface employed on many desktop computers which may provide for visual editing of the stored information either as a musical score, a table or event list of MIDI data, or in the form of a grid of some kind. Figure 5.18 shows a number of examples of different approaches to the display of stored MIDI information. Although it might be imagined that the musical score would be the best way of visualising MIDI data, it is often not the most appropriate. This is partly because unless the input is successfully quantised (see below) the score will represent *precisely* what was played when the music was recorded, and this is rarely good looking on a score! The appearance is often messy because some parts were slightly out of time, and the literal interpretation of what was played may be almost unrecognisable compared with what you thought you played. Score representation is useful after careful editing and quantisation, and can be used when you need to produce a visually satisfactory printed output. Alternatively, you might prefer to save the score as a MIDI file and export it to a music notation package for layout purposes.

In the grid editing display, notes may be dragged around using a mouse or trackball, and audible feedback is often available as the note is dragged up and down, allowing the user to hear the pitch or sound as the position changes. Note lengths may be changed by dragging their ends in or out, and the timing position may be altered by dragging the note left or right. In the event list form, each MIDI event is listed next to a time value. The information in the list may then be changed by typing in new times or new data values. Also events may be inserted and deleted. In all of these modes the familiar cut and paste techniques used in word processors and other software can be applied, allowing events to be used more than once in different places, repeated so many times over, and other such operations.

A whole range of semi-automatic editing functions are also possible, such as transposition of music, using the computer to operate on the data so as to modify it in a predetermined fashion before sending it out again. Transposition, for example, is simply a matter of raising or lowering the MIDI note numbers of every stored note by the relevant degree. You could also create echo effects by duplicating a track and offsetting it by a certain amount, for example. A sequencer's ability to search the stored data (both music and control) based on specific criteria, and to perform modifications or transformations to just the data which matches the search criteria, is one of the most powerful features of a modern system. For example, it may be possible to search for the highest-pitched notes of a polyphonic track so that they can be separated off to another track as a melody line. Alternatively it may be possible to apply the rhythm values of one track to the pitch values of another so as to create a new track, or to apply certain algorithmic manipulations to stored durations or pitches for compositional experimentation.

The possibilities for searching, altering and transforming stored data are almost endless once musical and control events are stored in the form of unique values, and for those who specialise in advanced composing or experimental music these features will be of particular importance. It is in this field that many of the high-end sequencer packages will continue to develop.

5.6.6 Quantisation of rhythm

Rhythmic quantisation is a feature of almost all sequencers, and is based upon the sequencer's ability to alter the timing of events. In its simplest form it involves the 'pulling-in' of note events to the nearest musical time interval at the resolution

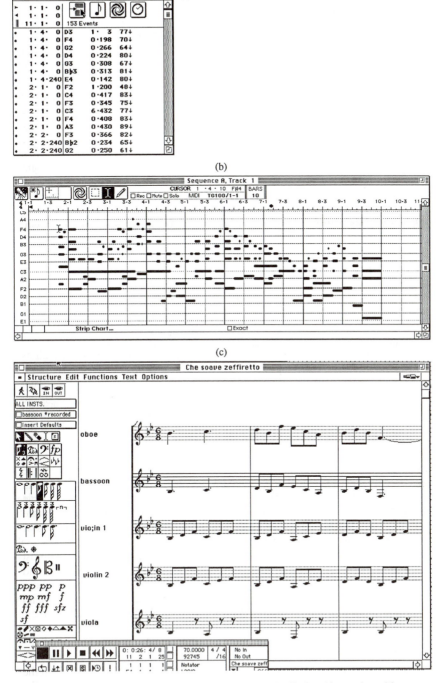

Figure 5.18 MIDI data, once stored in sequencer memory, can be displayed in a variety of forms. These are some examples: (a) event list (Studio Vision); (b) graphical form (Studio Vision); (c) musical score (Notator Logic)

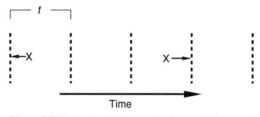

Time

Figure 5.19 In simple rhythm quantisation, notes (represented by X) are 'pulled in' to the nearest musical timing point (the minimum increment is represented by t). Notes over the halfway mark between one timing point and the next will be quantised to the next point. Advanced algorithms are available which allow the user to decide within what range out-of-time notes will be 'pulled in', and others make it possible to leave alone those notes which are only slightly out of time and quantise those more drastically adrift, thereby preserving the natural irregularity in playing whilst correcting obvious mistakes

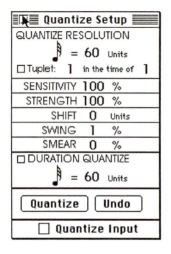

Figure 5.20 Quantisation options in Opcode Studio Vision. 'Sensitivity' determines how close a note has to be to the quantising interval (grid) to be quantised. If negative then further away notes are quantised and closer notes are left alone, whereas positive values have the opposite effect. 'Strength' determines how far notes are pulled towards quantising intervals; 'Shift' quantises notes and then shifts them out of time by so many units; 'Swing' moves every other quantising interval so that it is not exactly in between the two on either side; 'Smear' randomly moves notes after they have been quantised. If the duration is quantised then note durations are adjusted to the nearest specified increment

specified by the user, so that notes which were 'out of time' can be played back 'in time' (see Figure 5.19). It is normal to be able to program the quantisation resolution to an accuracy of at least as small as a 32nd note, and the choice depends on the audible effect desired. It must be borne in mind that notes played far enough out of time that they are over the halfway mark between one time interval and the next will go to the nearest time interval, which might be the one before or the one after that intended. In this case, editing would be required.

Events may be quantised either permanently or just on replay. The permanent method will alter the timing for ever after, whereas the replay-only quantisation may be changed at will and does not affect the stored data. Some systems allow 'record quantisation' which alters the timing of events as they arrive at the input to the sequencer. This is a form of permanent quantisation. It may also be possible to 'quantise' the cursor movement so that it can only drag events to predefined rhythmic divisions.

More complex rhythmic quantisation is also possible, in order to maintain the 'natural' feel of rhythm for example. Simple quantisation can result in music which sounds 'mechanical' and electronically produced, whereas the 'human feel' algorithms available in many packages attempt to quantise the rhythm strictly and then reapply some controlled randomness. The parameters of this process may be open to adjustment until the desired effect is achieved. Figure 5.20 shows an example of the controls provided by one package to control the nature of quantisation.

5.6.7 Arranging sequences

Complete sequences, often consisting of a number of tracks, may be edited together to form a longer song. For example, one might wish to create independent sequences corresponding to the main sections of a piece of music in order that a chorus could be repeated at certain junctures. A name or label of some kind will therefore be attached to each stored sequence, and each will have a defined length (normally at least the length of the longest track). The example in Figure 5.21 shows one way in which such completed sequences may be arranged by ordering the collection of labelled sequences graphically.

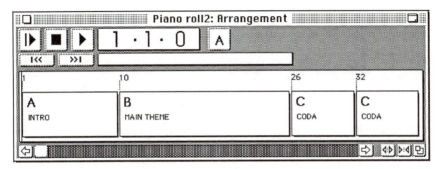

Figure 5.21 Sections or sub-sequences of a song may be edited together graphically by dragging the blocks appropriately as shown (Opcode EZ Vision). Sections may be repeated, lengthened or shortened in a similar way

5.6.8 Non-note events in MIDI sequences

In addition to note events, one may either have recorded or may wish to add events which are for other control purposes, such as program change messages, controller messages or system exclusive messages. If such messages are transmitted from the MIDI data source during recording then they will normally be stored as events alongside the note data (unless they have been filtered out). Such data may be displayed in a number of ways, but again the graphical plot is arguably the most useful. It is common to allow selected controller data to be plotted below the note data in a strip chart, such as shown in Figure 5.22. Here two examples are shown. The first one shows note velocities and the second shows movements of the volume controller.

It is possible to edit these events in a similar way to note events, but there are a number of other possibilities here. For example a scaling factor may be applied to controller data in order to change the overall effect by so many per cent, or a graphical contour may be drawn over the controller information to scale it according

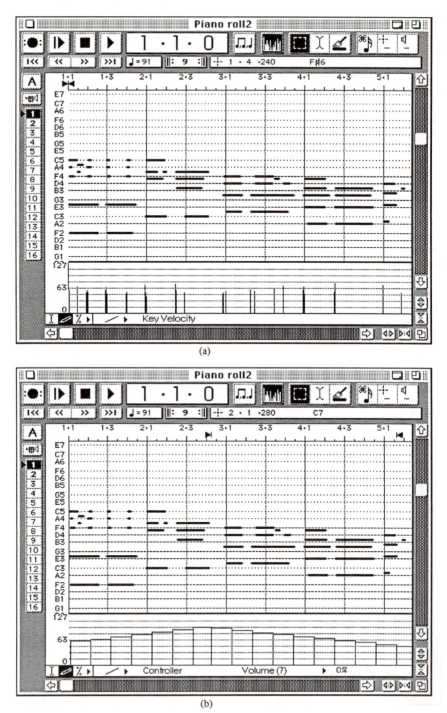

Figure 5.22 Two examples of control information displayed in a sequencer's strip chart.
(a) Note velocities, and (b) continuous controller data (e.g. MIDI volume). (Opcode EZ Vision)

to the magnitude of the contour at any point. Such a contour could be used to introduce a gradual increase in note velocities over a section, or to introduce any other time-varying effect.

Program changes may be inserted at any point in a sequence, usually either by inserting the message in the event list, or by drawing it at the appropriate point in the controller chart. This has the effect of switching the receiving device to a new voice or stored program at the point where the message is inserted. It can be used to ensure that all tracks in a sequence use the desired voices from the outset without having to set them up manually each time. Either the name of the program to be selected at that point or its number can be displayed, depending on whether the sequencer is subscribing to a known set of voice names such as General MIDI (see section 3.8.3), as shown in the example of Figure 5.23.

When editing continuous controller information it is often possible to apply a degree of 'thinning' to the messages. This limits the number of messages transmitted over the duration of the changing position of the control. Clearly too few messages could result in a jerky or step-like effect, but too many may be unnecessary for the purpose, may clog up the MIDI bus, and wastes memory. Experimentation may be required to determine a satisfactory degree of controller thinning.

System exclusive data may also be recorded or inserted into sequences in a similar way to the message types described above. Any such data received during recording will normally be stored and may be displayed in a list form. It is also possible to insert SysEx voice dumps into sequences in order that a device may be loaded with new parameters whilst a song is executing if required.

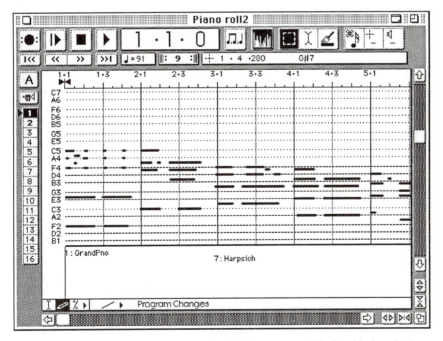

Figure 5.23 Program change messages can be stored in the sequence and displayed in the strip chart either by name or number. (Opcode EZ Vision)

5.6.9 Event chasing

When a sequence is replayed events are normally executed as they occur in particular musical positions. If you then cue to a point some way into the song and begin to play again, it may be that certain important events will have been skipped in the intervening bars, possibly causing the status of devices to be incorrect when replay begins again. For this reason, some sequencers may offer the option of 'chasing' events when cueing, which effectively means that the program will run through all the non-note events in each track, sending the events in quick succession to receiving devices in order that their status is correctly updated. Event chasing should only be used if it is important because it makes cueing quite a lot slower.

5.6.10 MIDI mixing and faders

Sequencers often provide a facility not for mixing audio, but for controlling the volume and panning of MIDI sound generators. This is an effective way of emulating audio mixing, especially if the sources concerned have their stereo outputs combined externally using a simple mixer. Using MIDI volume and pan controller numbers (decimal 7 and 10), a series of graphical faders can be used to control the audio output level of voices on each MIDI channel, and may be able to control the pan position of the source between the left and right outputs of the sound generator if it is a stereo source. On-screen faders may also be available to be assigned to other functions of the software, as a means of continuous graphical control over parameters such as tempo, or to vary certain MIDI continuous controllers in real time.

5.7 Some issues to be considered when comparing sequencers

With such a variety of software available it is often difficult to determine which package will be the most suitable for one's purposes. Often it is not possible for one package to fulfil all the requirements and you may end up using different software in different situations. Nonetheless there are some key points to be considered which may influence the decision:

Internal timing resolution
Although an external MIDI clock will only provide sync resolution of 24 pulses per quarter note (ppqn), most sequencers have internal resolutions which are considerably higher than this. By this it is meant that the resolution with which they time events is finer than 24 ppqn. It is not uncommon, for example, for commercial sequencer software to have internal resolutions of 480 or even 960 ppqn – this being the maximum timing accuracy to which events can be represented. When such sequencers are locked to an external clock (see Chapter 6) they will interpolate between received clock messages to achieve the required internal resolution. If very precise control over event timing is important for your application then this aspect of the specification should be carefully examined.

Balance between module features
As mentioned earlier, there is a grey dividing line between sequencers, notation packages and digital audio editing software. A number of high-end packages

provide some or all of these features, and it may be possible to buy the software in a modular form. If all that is required is to record and edit MIDI information, then it will be considerably cheaper to buy a package which does not include the latter two features, since music can be edited graphically in nearly all sequencers and a real musical score may not be necessary. On the other hand, if at the same time as recording a musical work you want to produce a printed score, then the notation features will become considerably more important, and it will be necessary to look for a package capable of producing good-looking printed music as well as offering comprehensive MIDI processing and editing facilities. True music notation packages produce the most professional-looking scores, but these often only have limited sequencer features, so the correct balance between these applications must be sought.

Digital audio may be recorded onto a computer hard disk alongside MIDI information (see section 5.9) and sequencer packages which include this option tend to be considerably more expensive than those which do not. It is nonetheless a valuable facility if acoustic instruments, sound effects, voices or speech are to be recorded alongside MIDI tracks, since the two types of information can be edited in parallel and the one software package can be used to produce a finished recording.

Presentation of information

Although it may seem a trivial point, the way in which a sequencer package presents information to the user, and the clarity of the control interface, are of vital importance to the usability of the system. There are some incredibly arcane user interfaces around, which hide very powerful features. Clearly a powerful package will take some time to learn, but the user interface should be designed so as to help rather than hinder the operator. Have a look at a number of packages before settling for one, in order to compare the immediate impressions of the display and controls. There is no point in buying a package that you will never be able to use properly.

Compatibility with hardware MIDI interface

As discussed earlier in this chapter, a large number of MIDI interfaces are available for desktop computers. Make sure that the software package that you choose is compatible with the MIDI interface that you will use. Sequencer software may come with software drivers to handle a range of interfaces, but some are more successful than others. Generally it is fairly certain that a sequencer manufacturer's own MIDI interface will perform with less trouble than a third party interface.

Compatibility with operating system and other software

Make sure that the sequencer is compatible with any operating system or MIDI extension that you wish to use on the computer. For example, an OMS-compatible sequencer will work much more smoothly with other OMS software. Although there are ways of forcing otherwise incompatible packages to work together, they are only to be attempted if you have plenty of time on your hands. Similarly, if it is intended to run more than one package at a time on the same computer, check that they are capable of operating in this fashion, and that they are both compatible with the same revision of system software.

Import and export facilities

If you need to exchange MIDI sequences with other users or with other packages, you will need a sequencer capable of importing and exporting its files in third party formats. Since the introduction of Standard MIDI Files (see section 5.8) the issue has been greatly simplified, since most sequencers are capable of reading and writing this common format, although it may be necessary for some parameters to be reset in the new package after importing a MIDI file.

Compatibility with voice librarians
It is useful if the sequencer is capable of interaction with a voice or patch librarian package, in order that banks of stored voice data can be addressed by name, in order to set program changes at different points in a song. It may also be useful if voice data can be downloaded to the MIDI devices from within the sequencer.

Synchronisation features
The sequencer's synchronisation features are important if you will need to lock replay to external timing information such as MIDI clock and timecode. Most recent sequencers are able to operate in either beat clock or timecode modes, and some may be able to detect which type of clock data is being received and switch over automatically. If you need to lock the sequencer to another sequencer or to a drum machine then beat clock synchronisation may be adequate. If you will be using the sequencer for applications which involve the timing of events in real rather than musical time, such as the dubbing of sounds to a film, then it is important that the sequencer is able to allow MIDI events to be tied to timecode locations as well as to musical beat locations, since timecode locations will remain in the same place even if the musical tempo is changed.

5.8 Standard MIDI files

Sequencers and notation packages typically store data on disk in their own unique file formats. Occasionally it is possible for a file from one package to be read by other, especially if the packages are from the same manufacturer, but this is rare. The standard MIDI file was developed in an attempt to make interchange of information between packages more straightforward, and is now used widely in the industry in addition to manufacturers' own file formats. It is rare now not to find a sequencer or notation package capable of importing and exporting standard MIDI files.

Three types of standard MIDI file exist to encourage the interchange of sequencer data between software packages. The MIDI file contains data representing events on individual sequencer tracks, as well as containing labels such as track names, instrument names and time signatures. It is the intention that not only should software packages running on the same computer type be able to read these universal files, but that such files may be ported to other computers (using one of the many file exchange protocols), so that a package running under a completely different operating system may read the data.

5.8.1 Exchanging MIDI files between systems

When exchanging MIDI files between systems there is the question of disk format compatibility to consider (see Figure 5.24). Although the MIDI data file itself is standardised, the medium on which it is stored is not. You could store a MIDI file on one of various densities of 3.5-inch floppy disk or on a CD-ROM, for example, and

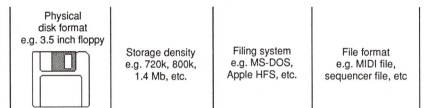

| Physical disk format e.g. 3.5 inch floppy | Storage density e.g. 720k, 800k, 1.4 Mb, etc. | Filing system e.g. MS-DOS, Apple HFS, etc. | File format e.g. MIDI file, sequencer file, etc |

Figure 5.24 A hierarchy of issues is concerned when exchanging disks between computers, as illustrated here

there are many ways in which such storage media may be formatted depending on the filing system in use by the host computer.

The first issue to be considered when attempting to transfer a MIDI file between systems, therefore, is reading the data from the mass storage medium. This requires using a drive which is compatible with the storage medium, and a host computer capable of reading the filing structure. A number of translation packages exist for this purpose, and these days it is a relatively straightforward matter to read files from a 'foreign' machine, using a multistandard disk drive. An alternative is to transfer the MIDI file over a network or point-to-point interface from a remote machine. In the desktop computer world, the Mac and MS-DOS file structures are the two most common at the moment, and Atari machines use the same basic data format on disk as MS-DOS (although compatibility is limited in practice), making the business of data exchange reasonably straightforward although not really simple. Most Macintosh disk drives, for example, are capable of reading and writing MS-DOS-formatted disks.

It is also possible to transfer MIDI files from one machine to another over a MIDI link, as discussed in section 5.8.8.

5.8.2 General structure of MIDI files

There are three MIDI file types. File type 0 is the simplest, and is used for single-track data, whilst file type 1 supports multiple tracks which are 'vertically' synchronous with each other (such as the parts of a song), and file type 2 contains multiple tracks which have no direct timing relationship and may thus be asynchronous. Type 2 could be used for transferring song files which are made up of a number of discrete sequences, each with a multiple track structure.

The basic file format consists of a number of 8 bit words formed into so-called 'chunks'. The header chunk, which always heads a MIDI file, contains global information relating to the whole file, whilst subsequent track chunks contain event data and labels relating to individual sequencer tracks. Track data should be distinguished from MIDI channel data, since a sequencer track may address more than one MIDI channel. Each chunk is preceded by a preamble of its own, which specifies the type of chunk (header or track) and the length of the chunk in terms of the number of data bytes which are to be contained in the chunk. There then follow the designated number of data bytes (see Figure 5.25).

As can be seen, the chunk preamble contains firstly a 4 byte section to identify the chunk type using the ASCII character format (four bytes equivalent to four characters), and secondly a 4 byte section (eight hexadecimal characters) to indicate

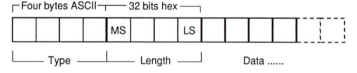

Figure 5.25 The general format of a MIDI file chunk. Each chunk has a preamble consisting of a 4 byte ASCII 'type' followed by 4 bytes to represent the number of data bytes in the rest of the message (the 'length')

the number of data bytes in the chunk (the length). The number of bytes indicated in this length message does not include the length of the preamble (which is always eight bytes).

5.8.3 Header chunk

The header chunk which begins every MIDI file takes the format shown in Figure 5.26. After the 8 byte preamble will normally be found 6 bytes containing header data, considered as three 16 bit words, the first of which ('format') defines the file type as 0, 1 or 2 (see above), the second of which ('ntrks') defines the number of track chunks in the file, and the third of which ('division') defines the timing format used in subsequent track events.

A zero in the MSB of the 'division' word indicates that events will be represented by 'musical' time increments of a certain number of 'ticks per quarter note' (the exact number is defined in the remaining bits of the word), whilst a one in the MSB indicates that events will be represented by real-time increments in number-of-ticks-per-timecode-frame. The frame rate of the timecode (see Chapter 6) is given in the remaining bits of the most significant byte of 'division', being represented using negative values in twos complement form. Thus the standard frame rates are represented by one of the decimal values –24, –25, –29 (for 30 drop frame) or –30. A negative number may be represented in twos complement form by taking its positive binary equivalent, inverting all the bits (zeros become ones and vice versa), then adding a binary one.

When a real-time format is specified (MSB of 'division' = 1) in the header chunk, the least significant byte of 'division' is used to specify the subdivisions of a frame to which events may be timed. For example, a value of '4_{10}' in this position would mean that events were timed to an accuracy of a quarter of a frame, corresponding to the arrival frequency of MIDI quarter-frame timecode messages, whilst a value of '80_{10}' would allow events to be timed to bit accuracy within the timecode frame (there are 80 bits representing a single timecode frame value in the SMPTE/EBU longitudinal timecode format).

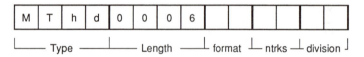

Figure 5.26 The header chunk has the type 'MThd' and the number of data bytes indicated in the 'length' is 6 (see text)

5.8.4 Track chunks

Following the header come a number of track chunks (see Figure 5.27), the number depending on the file type and the number of tracks. File type 0 represents a single track, and thus will only contain a header and one track chunk, whilst file types 1 and 2 may have many track chunks. Track chunks contain strings of MIDI events, each labelled with a delta-time at which the event is to occur. Delta-times represent the number of 'ticks' since the last event, as opposed to the absolute time since the beginning of a song. The exact time increment specified by a tick depends on the definition of a tick contained in the 'division' word of the header (see above).

Delta-time values are represented in 'variable length format', which is a means of representing hexadecimal numbers up to &0FFFFFFF as compactly as possible. Variable length values represent the number in question using one, two, three or four bytes, depending on the size of the number. Each byte of the variable length value has its MSB set to a one, except for the last byte whose MSB should be zero. (This distinguishes the last byte of the value from the others, so that the computer reading the data knows when to stop compiling the number.) There are thus seven bits of each byte available for the representation of numeric data (rather like the MIDI status and data bytes). A software routine must be written to convert normal hex values into this format and back again. The standard document gives some examples of hex numbers and their variable length equivalents, as shown in Table 5.1.

Table 5.1 Examples of numbers in variable length format

Original number (hex)	Variable length format (hex)
00000000	00
00000040	40
0000007F	7F
00000080	81 00
00002000	C0 00
00100000	C0 80 00
0FFFFFFF	FF FF FF 7F

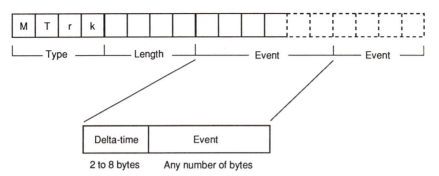

Figure 5.27 A track chunk has the type 'MTrk' and the number of data bytes indicated in the 'length' depends on the contents of the chunk. The data bytes which follow are grouped into events as shown

5.8.5 MIDI file track events

The track events which occur at specified delta-times fall into the categories of 'MIDI event', 'SysEx event' and 'meta-event'. In the case of the MIDI event, the data bytes which follow the delta-time are simply those of a MIDI channel message, with running status used if possible.

System exclusive (SysEx) events are used for holding MIDI system exclusive dumps that occur during a sequence. The event data is normally identical to the system exclusive data packet to be transmitted, except that the length of the packet is specified *after* the initial &[F0] byte which signals the beginning of a SysEx message, and before the normal manufacturer ID, as follows:

&[F0] [length] [SysEx data] … …

The 'length' value should be encoded in variable length format, and the standard requires that &[F7] be used to terminate a SysEx event in a MIDI file. (Some software omits this when transmitting such data over MIDI.)

It is also possible to have a special 'SysEx' event, as follows:

&[F7] [length] [data] … …

The standard says that this can be used as a form of 'escape' event, in order that data may be included in a standard MIDI file which would not normally be part of a sequencer file, such as real-time messages or MTC messages. The &F7 byte is also used as an identifier for subsequent parts of a system exclusive message which is to be transmitted in timed packets (some instruments require this). In such a case the first packet of the SysEx message uses the &F0 identifier and subsequent packets use the &F7 identifier, preceded by the appropriate delta-times to ensure correct timing of the packets.

The meta-event is used for such information as time signature, key signature, text, lyrics, instrument names and tempo markings. Its general format consists of a delta-time followed by the identifier &FF, as follows:

Table 5.2 A selection of common meta-event type identifiers

Type (hex)	Length	Description
00	02	Sequence number
01	Var	Text event
02	Var	Copyright notice text
03	Var	Sequence or track name
04	Var	Instrument name
05	Var	Lyric text (normally one syllable per event)
06	Var	Marker text (rehearsal letters, etc.)
07	Var	Cue point text
20	01	MIDI channel prefix (ties subsequent events to a particular channel, until the next channel event, MIDI or meta)
2F	00	End of track. (No data follows)
51	03	Set tempo (μs per quarter note)
54	05	Timecode location (hh:mm:ss:ff:100ths) of track start (following the MTC convention for hours, see Chapter 6)
58	04	Time signature (see below)
59	02	Key signature. First data byte denotes number of sharps (+ve value) or flats (–ve value). Second data byte denotes major (0) or minor (1) key
7F	Var	Sequencer-specific meta-event (see above)

&[FF] [type] [length] [data] … …

The byte following &FF defines the type of meta-event, and the 'length' value is a variable length number describing the number of data bytes in the message which follows it. The number of bytes taken up by 'length' therefore depends on the message length to be represented.

Many meta-events exist, and it is not intended to describe them all here, although some of the most common type identifiers are listed in Table 5.2. A full list of current meta-events can be obtained from the IMA. It is allowable for a manufacturer to include meta-events specific to a particular software package in a MIDI file, although this is only recommended if the standard MIDI file is to be used as the *normal* storage format by the software. In such a case the 'type' identifier should be set to &7F. A software package should expect to encounter events in MIDI files which it cannot deal with, and be able simply to ignore them, since either new event types may be defined after a package has been written, or a particular feature may be unimplemented.

5.8.6 Time signature format

The format of time signature meta-events needs further explanation, since it is somewhat arcane. The event consists of four data bytes following the 'length' identifier, as shown in Figure 5.28. The first two of these define the conventional time signature (e.g.: 4/4 or 6/8) and the second two define the relationship between MIDI clocks and the notated music.

The denominator of the time signature is represented as the power of two required to produce the number concerned. For example, this value would be &03 if the denominator was 8_{10} because 2^3 equals 8. The third data byte defines the number of MIDI clocks per metronome click (the metronome may click at intervals other than a quarter note, depending on the time signature), and the final byte allows the user to define the number of 32nd notes actually notated per 24 MIDI clocks. This last, perhaps unusual-sounding definition allows for a redefinition of the tempo unit represented by MIDI clocks (which would normally run at a rate of 6 per 16th note), in order to accommodate software packages which allow this relationship to be altered.

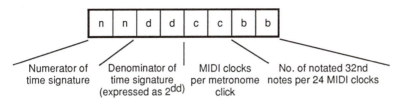

Figure 5.28 Meaning of the data bytes in the time signature meta-event

5.8.7 Tempo maps

The tempo map of a song may need to be transferred between one machine and another, and the MIDI file format may be used for this purpose. Such a file could be a type 0 file consisting solely of meta-events describing tempo changes, but otherwise the map must be contained in the first track chunk of a larger file. This is where reading devices will expect to find it.

5.8.8 Transferring files over a MIDI link

A relatively recent group of universal non-realtime messages has been set aside for the purpose of transferring files between machines using MIDI (they can be MIDI files or indeed any other type of file). As discussed elsewhere, universal non-realtime messages are system exclusive messages intended for the transfer of information which is not manufacturer specific, and which is not time critical.

The file dump messages have a sub-ID #1 of &07, and are almost identical in structure to the universal sample dump format (see section 3.14.2), and readers should refer to this section for details. Unlike the sample dump, the file dump uses both sub-ID #1 and #2 for its messages (although not for handshaking), whereas most sample dump messages only use a single sub-ID (see section 2.5.13.) Like the sample dump, transfer is possible using either open or closed loops, the closed loop allowing for the receiver to control the transfer using handshaking. The same handshaking messages are used as for sample dumps (i.e.: WAIT, CANCEL, ACK and NAK), with the addition of an EOF (end of file) identifier having the value &7B. It is not possible to describe the complete standard here and readers should also refer to the 4.2 Addendum to the MIDI 1.0 specification for further details (see Appendix 2).

The general format of file dump messages is different from the sample dump in that it specifies a source device ID as well as a destination ID. The source ID byte follows the two sub-IDs, as follows:

&[F0] [7E] [dest. ID] [07] [sub-ID #2] [source ID] [data] ... ... [F7]

This allows the source of a file dump request to identify itself, in order that the receiver knows which device to respond to with handshaking messages. The source ID is only normally sent in the file dump header, and is not included in the transmission of subsequent packets. The sub-ID #2 identifies the message as either a header (&01), a data packet (&02) or a dump request (&03).

A dump request incorporates the following [data] within the general format shown above:

[type] [type] [type] [type] [name] [name] ... ...

where the [type] bytes are four 7 bit ASCII characters describing the file type (e.g.: 'MIDI'), and the [name] bytes are as many ASCII characters as required to spell out the file name (this is optional).

A dump header describes the vital statistics of the file and incorporates the following [data] within the general format:

[type] [type] [type] [type] [length] [length] [length] [length] [name] [name] ... ...

where [type] and [name] are as above, and the [length] bytes (transmitted LSB first) create a 28 bit word describing the size of the file in bytes.

Data packets are similar to sample dump packets (except that they include both sub-IDs #1 and #2), and, in order to accommodate the transfer of 8 bit values over MIDI, eight message bytes are used to transfer seven actual bytes of the file, since MIDI data bytes only carry seven active bits. The general format of a packet is as follows:

&[F0] [7E] [dest. ID] [07] [02] [packet no.] [byte count] [data] ... ... [LL] [F7]

where the packet number increments as with sample dumps, and the byte count is the number of encoded data bytes in the packet, minus one (maximum 128 data bytes in the packet). Seven data bytes of the file are encoded as eight bytes by first transmitting the 7 MSBs together and in order as one MIDI byte, followed by the remaining LSBs of each data byte in one MIDI byte each. LL is a checksum, as for sample dumps.

5.9 Integrating digital audio and video

As mentioned earlier, digital audio recording is becoming a popular feature of high-end sequencer packages. Digital video is also an increasingly viable option. Using either the internal signal processing of the computer or an add-on card, digital audio and video may be recorded on a disk drive and synchronised with sequenced MIDI data. The MIDI tracks could be used to store control information for electronic instruments, the digital audio tracks could be used to store 'real' sound signals such as vocals, speech, effects, and live instrumental parts, whilst a digital video cue track could be stored for the purpose of dubbing music or effects to picture. The term 'non-linear recording' is often used to describe the storage of audio and video in a random access form on computer mass storage media.

In this section of the book a short overview will be given, showing how digital audio and video recording may be integrated with MIDI sequencing. It is beyond the scope of this book to cover these subjects in detail, and the reader is referred particularly to two books by the same author which describe the principles and operation of digital audio systems in more detail. These are *Digital Audio Operations* and *Tapeless Sound Recording*, listed in Appendix 2 at the end of this book.

5.9.1 Adding digital audio capability

If digital audio capability is to be added to a desktop computer, with the intention of running it alongside MIDI sequencing, it will normally be necessary to install a third party expansion card. There is also the possibility that a recent computer may have sufficient internal signal processing to handle digital audio recording without the need for extra cards.

The basic principles of digital audio sampling and conversion were described in section 3.14 on MIDI samplers, and the same processes apply here. Figure 5.29 shows a typical block diagram of the elements involved. Audio is converted to the digital domain using an A/D convertor, often mounted externally in a rack unit, or it may be derived directly from a digital source using one of the standard digital audio interfaces such as AES/EBU or SPDIF. (For detailed coverage of digital interfacing the reader is referred to *The Digital Interface Handbook*, as listed in Appendix 2.) Audio is normally stored on a fast disk drive such as a Winchester drive or removable optical drive, and this is normally interfaced to the computer or directly to the audio expansion card using SCSI (the Small Computer Systems Interface).

Assuming the use of linear PCM digital audio, the two most common sampling rates in use are 44.1 and 48 kHz, and the majority of current systems use 16 bit quantisation (although there is a gradual trend towards 20 or 24 bits). Such sampling

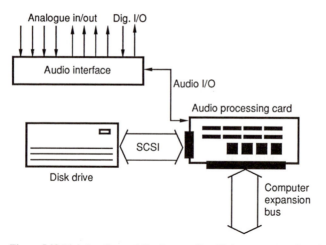

Analogue in/out   Dig. I/O

Audio interface

Audio I/O

Audio processing card

SCSI

Disk drive

Computer expansion bus

Figure 5.29 Digital audio capability is normally added to a computer-based MIDI system using an audio expansion card. This card communicates with an external audio interface in which are housed A/D and D/A convertors, as well as digital audio interfaces. Audio data is stored on a large disk drive, interfaced using SCSI. The expansion card houses audio buffer memory and DSP devices for audio signal processing. It also communicates with the rest of the computer via the computer's internal buses

parameters mean that digital audio typically occupies a great deal more disk space than the equivalent duration of an accompanying MIDI sequence, often by a factor of many hundreds. For example, one hour of single channel audio at 48 kHz, 16 bits, requires around 330 Mbytes of disk storage. There is a pro rata increase if the number of channels or the amount of storage time is increased. It follows that one normally requires a much larger disk drive when digital audio capability is added to a MIDI system. (If the system uses some form of audio data reduction then it is possible that the storage requirement may be reduced by a factor of up to ten, but the effects of data reduction on sound quality should be carefully assessed before opting for this apparently beneficial solution.)

Digital audio is stored on a computer disk as a file. The disk is formatted into tracks and sectors, as shown in Figure 5.30, and may have a number of physical surfaces if it is a Winchester hard disk. A typical sound file will occupy many blocks on the disk, and these blocks may not be contiguous (that is they may not be located adjacent to each other). For this reason a certain amount of RAM buffering is required, either within the computer or on the expansion card, to smooth the transfer of audio to and from the drive. An unbroken stream of audio is present at the digital audio inputs and outputs, but the flow to and from the disk will be erratic. The situation with memory buffering may be likened to a bucket being filled with water from a tap, as shown in Figure 5.31. Provided that the bucket (analogous to the memory buffer) is kept reasonably full, water (analogous to audio) will continue to flow in an unbroken stream from the hole in the bottom, even if the bucket is filled in bursts from the tap. Similarly, continuous filling could be translated into burst emptying. The only important criteria are (a) that the average rate of flow of water into and out of the bucket are the same, and (b) that the size of the bucket is adequate to accommodate the discontinuities in flow.

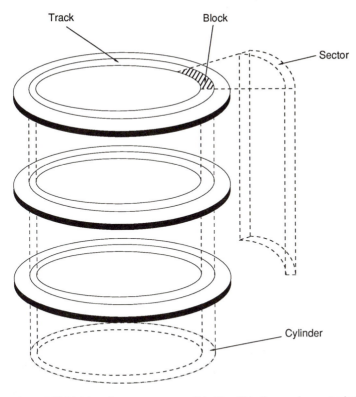

Track   Block   Sector

Cylinder

Figure 5.30 Division of storage space on a disk drive. This diagram shows a typical Winchester hard disk with multiple surfaces, in which case the three-dimensional equivalents of tracks and sectors are blocks and cylinders. Optical disks only have one platter, are read one side at a time and the track is normally a continuous spiral, thus only the sector concept applies

Disk drives for digital audio use must meet certain basic performance criteria, which depend on the system concerned and the number of channels it is designed to handle simultaneously. These criteria relate to the transfer rate and access time of the drive. As shown in Figure 5.32, the access time is made up of the 'seek' latency (the time taken for the head to move to the required track on the disk), and rotational latency (the time taken for the required block to come under the head). The sustained transfer rate is normally more important than the so-called 'burst' transfer rate, since this represents more closely what happens in real operation. Access times of less than 10 ms are now very common, as are transfer rates of many tens of megabits per second. A single channel of 48 kHz, 16 bit audio requires a transfer rate of just under one megabit per second. It is not possible to specify precisely what performance is required from a disk drive, since this varies from system to system, and manufacturers often insist on the adoption of one of a small selection of approved drives which conform to their requirements.

When adding digital audio capability to a MIDI system one should consider the potential for expandability of the hardware. Recent computers with internal high speed digital signal processing (DSP) may be able to handle the recording of two or

(a)

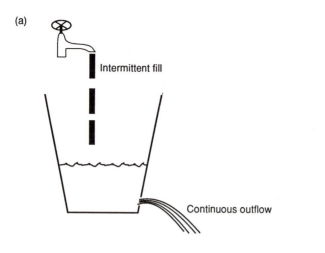

Intermittent fill

Continuous outflow

(b)

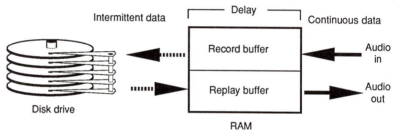

Intermittent data   Delay   Continuous data

Record buffer   Audio in

Replay buffer   Audio out

Disk drive

RAM

Figure 5.31 The bucket of water shown in (a) is likened to a memory buffer which acts to translate burst flow of data from a disk into continuous flow, or vice versa, as shown in (b)

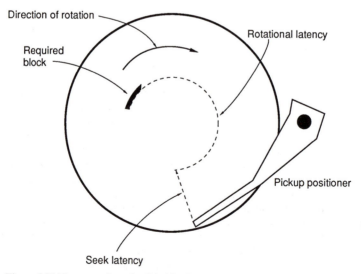

Direction of rotation

Rotational latency

Required block

Pickup positioner

Seek latency

Figure 5.32 The access time of a disk drive is made up of seek and rotational latency

four channels of digital audio without the need for add-on cards. Depending on the software chosen, it may be possible to expand this channel capacity through the use of third party cards, but this is not automatically to be assumed. A number of the third party products are designed with modularity in mind, allowing the user to install additional boards in chunks of, say, four channels at a time, as needs develop. Digidesign, one of the leading manufacturers of low cost digital hardware and software, produces an expansion chassis, for example, which allows more audio cards to be installed than the average computer can hold internally.

When attempting to decide how many channels of digital audio are required it is worth bearing in mind that one can usually manage with many fewer simultaneous channels than would normally be required for multitrack recording in the tape-based sense. Firstly, much of the material will be played in real time by the MIDI sequencer and need not necessarily be recorded on the digital audio tracks. Secondly, most packages allow for a much larger number of *tracks* to be assembled than there are simultaneous outputs. The only limitation is on the number of sounds which can be played together. Thirdly, there is almost unlimited capacity for 'bouncing-down' of multiple audio tracks to create new files which are mixed versions of existing tracks. Since this bouncing is in the digital domain there is no loss of quality as is normally the case with analogue tape recording.

5.9.2 Digital audio software

Many of the leading sequencers optionally offer digital audio capabilities. Typically, such software requires the presence of one of the common third party expansion cards or the internal audio processing capability described in the previous section. For example, nearly all the Macintosh-based sequencers which offer digital audio require the installation of at least one Digidesign audio card such as 'ProTools', 'Sound Tools' or 'AudioMedia II'. It is very convenient operationally when common digital audio hardware is used by multiple software packages, since it makes possible easy switching between applications and only requires the purchase of one audio expansion kit.

Integrated MIDI and digital audio information is presented to the user in a number of different ways, and an example is shown in Figure 5.33. Tracks are usually depicted horizontally, either as a detailed waveform display of the audio signal which may be zoomed in to various levels of detail to allow precise editing, or as named blocks occupying a certain duration. Sound files may be broken up into smaller sub-sections and named independently, and any file or a sub-section of it may be used in a number of different locations and tracks without the need to duplicate the audio data. Crossfades may be introduced where one audio segment joins another, and the length and shape of the crossfade may normally be altered. When integrated with MIDI sequencing it may be possible to quantise digital audio events rhythmically, rather as MIDI events can be quantised. This is achieved in one package by allowing the user to strip out the silence between audio events (say the beats of a drum) by removing everything below a certain audio threshold. Thereafter each beat of the drum is a separate audio 'entity' and its start point may be rhythmically quantised. Many of the other arrangement features found in sequencers can also be applied to the digital audio tracks. In this way, MIDI control data and digital audio data may be edited and arranged in parallel.

Most software of this type also includes signal processing and automation features. It is normally possible at least to control the volume and panning of each

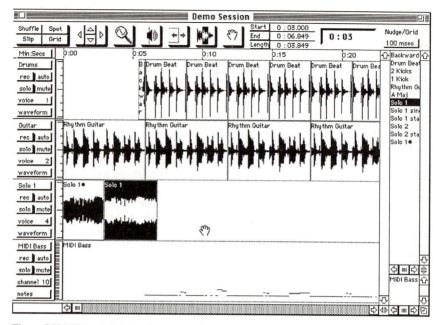

Figure 5.33 MIDI and digital audio data may be displayed and edited alongside each other, as shown in this example from Digidesign's ProTools package

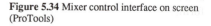

Figure 5.34 Mixer control interface on screen (ProTools)

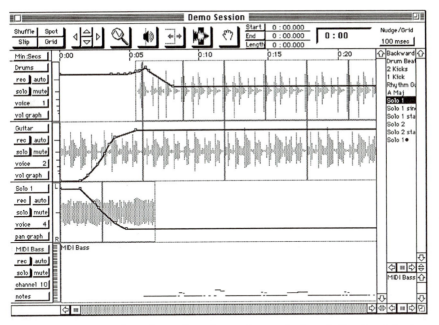

Figure 5.35 Audio processing options such as volume and pan can be dynamically automated and displayed in the form of a gain 'contour' as shown in this example from ProTools. Track 1 shows a volume contour, and Track 2 shows a pan contour

of the audio tracks, and it may be possible to apply other audio processing such as equalisation and effects. Such processing may be automated, often using a familiar 'mixer-style' user interface as shown in Figure 5.34, such that changes are stored with respect to time. The results may be displayed graphically using contour lines overlaid on the audio waveform, as shown in the example in Figure 5.35.

5.9.3 Digital audio files

The standard MIDI file is used widely for exchanging MIDI data between sequencers from different manufacturers, but unfortunately there is no directly comparable equivalent for digital audio purposes as yet. The American company Avid has recently introduced the Open Media Framework Interchange (OMFI), which is an open standard for the exchange of audio, video and edit list information, and a large number of companies have signed up (see Appendix 2). It is possible, therefore, that OMFI files will emerge as a standard for audio interchange. In the meantime, since Digidesign's sound cards have been installed more widely in desktop computers for MIDI-integrated digital audio recording than any other third party sound card, the Sound Designer 1 and 2 file formats used by Digidesign are often used as a common interchange format. Apple's AIFF (Audio Interchange File Format) is also offered as an option by a number of systems, as a means of exporting digital audio to a file format compatible with other packages.

5.9.4 Adding digital video capability

Recent advances in data reduction technology have brought digital video capability within the economic reach of desktop production systems. It is now possible to store and replay full motion video on a desktop computer, either using a separate monitor or within a window on an existing monitor. The replay of video from disk can be synchronised to the replay of digital audio and MIDI, using timecode, and this is particularly useful as a replacement for video on a separate video tape recorder (which is mechanically much slower, especially in locating distant cues).

Digital video at professional sampling rates and resolutions is particularly greedy when it comes to storage requirements – much more so than digital audio. For example, broadcast component digital video generates data at a rate of 270 Mbit/s, which would fill a typical hard disk in seconds even if the disk drive could handle the data rate. For this reason it is only practical to use digital video on today's desktop computers if its data rate is reduced drastically, and chips are now available which perform this operation in real time, reducing the data rate of pictures by over a hundred times to rates nearer that of linear digital audio. There is a consequent reduction in picture quality which depends on the amount of data reduction (although the quality can still be surprisingly good), and in any case even the poorest pictures stand comparison with VHS video (which is what many people were using before). The main advantage, as with disk-stored digital audio, is in the ability to locate instantaneously to any point in the program, without waiting for tape machines to park and cue up.

In the applications which we are considering here, compressed digital video is intended principally as a cue picture which can be used for writing music or dubbing sound to picture in post-production environments. In such cases the picture quality must be adequate to be able to see cue points, and possibly lip sync, but it does not need to be of professional broadcast quality. What is important is reasonably good slow motion and freeze-frame quality, and this requires the use of video data reduction techniques which code each frame individually, rather than coding frame differences or relying on inter-frame prediction. Individual frame coding is one of the key features of the JPEG (Joint Picture Experts Group) compression schemes, and a number of third party boards are available offering this type of compression. There are also a number of proprietary schemes in existence which fulfil broadly similar purposes.

5.10 MIDI development software

A lot of sequencer packages and other MIDI software is written in high level languages such as 'C' and Pascal. Time-critical parts of packages are often written in assembler or machine code. Programming in these languages requires considerable experience and is outside the intended scope of this book, although there are a number of books which cover the topic in detail (see Appendix 2). There are, however, a few tools which open up the world of MIDI programming to a wider range of potential users, taking advantage of object-orientated environments.

A good example of such a package is Opcode's 'MAX', named after one of the fathers of electronic music, Max Matthews. MAX allows functional objects to be linked graphically by dragging 'wires' from outputs to inputs, making it possible for the user to construct virtually any MIDI control 'engine' out of the building blocks.

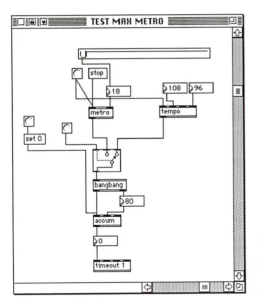

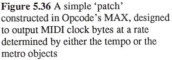

Figure 5.36 A simple 'patch' constructed in Opcode's MAX, designed to output MIDI clock bytes at a rate determined by either the tempo or the metro objects

MAX is essentially a MIDI construction kit, but it also allows new objects to be written in 'C' for those whose ambitions extend further than the built-in functions. A worldwide network of MAX developers exists, and a large number of third party objects have been authored, many of which are available for the asking. An example of a MAX program or 'patch' is shown in Figure 5.36. This simple program sends MIDI clock bytes to a predefined output port at a rate determined by the 'tempo' or 'metro' objects.

A program called 'HyperMIDI' is also useful for developing one's own MIDI applications, and it runs under Apple's Hypercard programming environment.

5.11 Music notation issues

5.11.1 Software packages

A number of music notation packages exist which are distinct from sequencers because they concentrate on producing a professional-looking printed score. Nonetheless they have certain sequencing features and they can normally read and write standard MIDI files which makes exchanging data with sequencers very easy. Scores can be produced either by dragging notes from a palette to the stave, by loading data from a MIDI file, or by entering pitches from a MIDI keyboard whist selecting note durations from either the palette or the QWERTY keyboard of the computer. Like a sequencer, individual lines of the score can be transmitted on predetermined MIDI channels in order that the score can be replayed using a sound generator.

Individual instrumental parts can then be separated from the full score, and comprehensive editing features are available to allow full control over the printed appearance of the score, with all of the normal musical notation and editorial marks

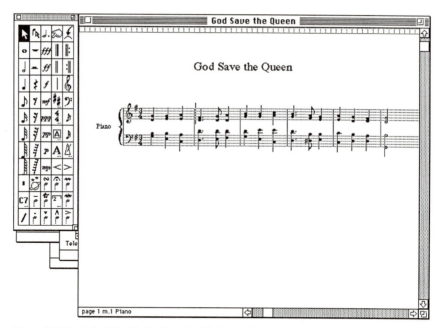

Figure 5.37 Typical editing display from the 'Nightingale' music notation package (Temporal Acuity Products), showing the palette of tools on the left-hand side

available for use. The example in Figure 5.37 shows the editing screen from a recently released notation package called 'Nightingale', showing the symbol palette and the score display.

5.11.2 MIDI notation messages

A group of universal realtime SysEx messages has been defined for transmitting information more closely related to the musical score than the majority of MIDI messages, using the sub-ID #1 of &03. The 'Time Signature Immediate' message behaves in the same way as its equivalent in the standard MIDI file (see section 5.8.6), whereas the 'Time Signature Delayed' message is only implemented at the next 'Bar Marker'. The bar marker message is intended to extend synchronisation possibilities by making it possible to define where bars begin.

Bar markers take the following general form:

&[F0] [7F] [dev. ID] [03] [01] [aa] [aa] [F7]

where [aa] [aa] are a pair of data bytes making up a 14 bit value (the MSB of each doesn't count), covering a positive and negative range using the twos complement convention, as shown in Figure 5.38. The LSbyte is transmitted first.

In normal operation bar numbers count up from zero, with bar 1 being the first bar, but it is possible to count in to the first bar by incrementing negative numbers up to zero. (Bar zero is effectively the last bar of count-in). The largest negative bar number (transmitted as &[00] [40] because the LSbyte is first) is used to signify that the bar count is temporarily not running, whilst the largest positive number

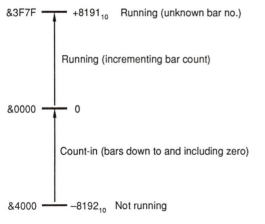

&3F7F ——┬—— $+8191_{10}$ Running (unknown bar no.)

Running (incrementing bar count)

&0000 ——┬—— 0

Count-in (bars down to and including zero)

&4000 ——┬—— -8192_{10} Not running

Figure 5.38 Diagram showing the ranges and functions of bar marker numbers

(transmitted as &[7F] [3F]) is used to indicate that the count is running but without a known bar number.

The bar marker normally takes effect either at the next MIDI clock byte or the next quarter-frame timecode message, but it is not intended to replace either or these as a means of low level synchronisation. It should normally be transmitted between the last MIDI clock of one bar and the first clock of the next. It is different from the song pointer, though, because the song pointer is really a means of autolocation rather than a means of high level synchronisation. The bar marker is a real-time message, which gives it a high priority, whereas the song pointer is not a real-time message. If bar markers are transmitted by the master timing device it is possible for a slave to find its place during externally synchronised replay without having to keep a count of the number of MIDI clocks elapsed.

The time signature messages take the general form:

&[F0] [7F] [dev. ID] [03] [02 or 42] [length] [data] [data] … …

where the data bytes follow the format of the time signature meta-event in the standard MIDI file (see section 5.8.6 above), except that it is possible to add additional groups of data bytes at the end of the message to signal the use of more than one time signature in a bar. The delayed version of the time signature is signalled with the sub-ID #2 of &42.

Chapter 6

Synchronisation and machine control

6.1 Introduction

An important aspect of MIDI control is the handling of timing and synchronisation data. MIDI timing data takes the place of the various older standards for synchronisation on drum machines and sequencers which used separate 'sync' connections carrying a clock signal at one of a number of rates, usually described in pulses-per-quarter-note (ppqn). There used to be a considerable market for devices to convert clock signals from one rate to another, so that one manufacturer's drum machine could lock to another's sequencer, but MIDI has supplanted these by specifying standard synchronisation data which may share the same data stream as the note and control information described in previous chapters.

Not all devices in a MIDI system will need access to timing information – it depends on the function fulfilled by each device. A sequencer, for example, will need some speed reference to control the rate at which recorded information is replayed, and this speed reference could either be internal to the computer or it might be provided by an external device. Drum machines should also take notice of timing information, but that is because a drum machine usually contains a sequencer which stores the patterns of rhythm which a player may lay down. On the other hand, a normal synthesiser, effects unit or sampler is not normally concerned with timing information, because it has no functions which would be affected by a timing clock. Such devices do not normally store rhythm patterns, although there are some keyboards with onboard sequencers which ought to recognise timing data.

As MIDI equipment has become more integrated with audio and video systems the need has arisen to incorporate timecode handling into the standard and into software. This has allowed sequencers either to operate relative to musical time (e.g.: bars and beats) or to 'real' time (e.g.: minutes and seconds). Using timecode, MIDI applications may be run in sync with the replay of a tape recorder, in order that the long-term speed relationship between the MIDI replay and the tape recorder remains constant. Also relevant to the systems integrator is the MIDI Machine Control standard which specifies a protocol for the remote control of devices such as tape recorders using a MIDI interface. This chapter, therefore, is concerned with explaining the various provisions made in MIDI by which MIDI devices may be synchronised with each other and with non-MIDI studio equipment, as well as examining the use of MIDI for remote machine control. Practical systems involving synchronisation are described in Chapter 7.

6.2 Music-related timing data

This section describes the group of MIDI messages which deals with 'music-related' synchronisation – that is synchronisation related to the passing of bars and beats as opposed to 'real' time in hours, minutes and seconds. It is normally possible to choose which type of sync data is used by a software package or other MIDI receiver when it is set to 'external sync' mode.

6.2.1 System realtime messages

A group of system messages called the 'system realtime' messages control the execution of timed sequences in a MIDI system, and these are often used in conjunction with the *song pointer* (which is really a system common message) to control autolocation within a stored song. The system realtime messages concerned with synchronisation, all of which are single bytes, are:

&F8	Timing clock
&FA	Start
&FB	Continue
&FC	Stop

The timing clock (often referred to as 'MIDI beat clock') is a single status byte (&F8) to be issued by the controlling device six times per MIDI beat. A MIDI beat is equivalent to a musical semiquaver or sixteenth note (see Table 6.1) so the increment of time represented by a MIDI clock byte is related to the duration of a particular *musical* value, not directly to a unit of real time. 24 MIDI clocks are therefore transmitted per quarter note, unless the definition is changed. (As has been seen in the discussion of time signature format in MIDI files (see section 5.8.6) some software packages allow the user to redefine the notated musical increment represented by MIDI clocks.) At any one musical tempo, a MIDI beat could be said to represent a fixed increment of time, but this time increment would change if the tempo changed.

The timing clock byte, like other system realtime messages, may temporarily interrupt other MIDI messages, with the status reverting to the previous status automatically after the realtime message has been handled by a receiver. This is necessary because of the very nature of the timing clock as a synchronising message. If it were made to wait for other messages to finish, it would lose its ability to represent a true increment of time. It may be seen that there could still be a small amount of error in the timing of any clock byte within the data stream if a large amount of other data was present, because the timing byte may not interrupt until at least the break between one byte and another, but this timing error cannot be greater than plus or minus half the duration of a MIDI byte, which is 160 μs.

Table 6.1 Musical durations related to MIDI timing data

Note value	Number of MIDI beats	Number of MIDI clocks
Semibreve (whole note)	16	96
Minim (half note)	8	48
Crotchet (quarter note)	4	24
Quaver (eighth note)	2	12
Semiquaver (sixteenth note)	1	6

So the &F8 byte might appear between the two data bytes of a note on message, for example, but it would not be necessary to repeat either the entire message or the 'note on' status after &F8 had passed. &F8 may also interrupt running status in the same way, without the need for reiteration of the status after the timing byte has been received. MIDI clocks should be given a very high priority by receiving software, since the degree of latency in the handling of this data will affect the timing stability of synchronised replay. On receipt of &F8, a device which handles timing information should increment its internal clock by the relevant amount. This in turn will increment the internal song pointer after six MIDI clocks (i.e. one MIDI beat) have passed. Any device controlling the sequencing of other instruments should generate clock bytes at the appropriate intervals, and any changes of tempo within the system should be reflected in a change in the rate of MIDI clocks. In systems where continuously varying changes have been made in the tempo, perhaps to imitate *rubato* effects or to add 'human feel' to the music, the rate of the clock bytes will reflect this.

The 'start', 'stop' and 'continue' messages are used to remotely control the receiver's replay. 'Start' is used by a controller to signal the start of replay of a prerecorded song. It should result in the playback of the sequence from the very beginning. 'Stop' is used to halt the replay of a song which is running on a synchronised device, and 'continue' is used to restart the replay from the point at which the sequence was last stopped or located to, not necessarily from the beginning. A receiver should only begin to increment its internal clock or song pointer after it receives a start or continue message, even though some devices may continue to transmit MIDI clock bytes in the intervening periods. For example, a sequencer may be controlling a number of keyboards, but it may also be linked to a drum machine which is playing back an internally stored sequence. The two need to be locked together, and thus the sequencer (running in internal sync mode) would send the drum machine (running in external sync mode) a 'start' message at the beginning of the song, followed by MIDI clocks at the correct intervals thereafter to keep the timing between the two devices correctly related. If the sequencer was stopped it would send 'stop' to the drum machine, whereafter 'continue' would carry on playing from the stopped position, and 'start' would restart at the beginning. Further practical examples of synchronised situations are given later in this chapter.

This method of synchronisation appears to be fairly basic, as it allows only for two options: playing the song from the beginning or playing it from where it has been stopped. It is difficult, if not impossible, with this method to start the song from a point in the middle. Originally this was the only system available, until the introduction of song pointers.

6.2.2 Song position pointers (SPPs)

SPPs are used when one device needs to tell another where it is in a song. (The term 'song' is used widely in MIDI parlance to refer to any stored sequence.) A sequencer or synchroniser should be able to transmit song pointers to other synchronisable devices when a new location is required or detected. For example, one might 'fast-forward' through a song and start again twenty bars later, in which case the other timed devices in the system would have to know where to restart. An SPP would be sent followed by 'continue' and then regular clocks. Originally it was recommended that a gap of at least 5 seconds was left between sending a SPP and restarting the sequence, in order to give the receiver time to locate to the new position, but recent revisions state that a receiver should be able to register a 'continue' message and count subsequent MIDI clocks even while still locating, even if it is not possible to start playing immediately. Replay should begin as soon as possible, taking into account the clocks elapsed since the 'continue' message was received.

An SPP represents the position in a stored song in terms of number of MIDI beats (not clocks) from the start of the song. Thus again it relates directly to musical time rather than real time, and it does not necessarily represent the number of seconds into the song. It uses two data bytes, and the MSB of data bytes must be zero, so the resulting fourteen bits can specify up to 16 384 MIDI beats. SPP is a system common message, not a realtime message. It is often used in conjunction with &F3 (song select), which is used to define which of a collection of stored song sequences (in a drum machine, say) is to be replayed.

SPPs are fine for directing the movements of an entirely musical system, in which every action is related to a particular beat or subdivision of a beat, but not so fine when actions must occur at a particular point in real time. If, for example, one was using a MIDI system to dub music and effects to a picture in which an effect was intended to occur at a particular visual event, that effect would have to maintain its position in time no matter what happened to the music. If the effect was to be triggered by a sequencer at a particular number of beats from the beginning of the song, this point could change in real time if the tempo of the music was altered slightly to fit a particular visual scene. Clearly some means of real-time synchronisation is required either instead of, or as well as the clock and song pointer arrangement, such that certain events in a MIDI controlled system may be triggered at specific *times* in hours, minutes and seconds.

6.2.3 Other musical synchronisation messages

Recent software may recognise and be able to generate the bar marker and time signature messages. These universal realtime messages were described in section 5.11.2. The bar marker message can be used where it is necessary to indicate the point at which the next musical bar begins, and normally takes effect at the next &F8 clock.

6.2.4 Synchronisation to an audio click or tap

Some MIDI synchronisers will accept an audio input or a tap switch input so that the user can program a tempo track for a sequencer based on the rate of a drum beat or a rate tapped in using a switch. This can be very useful in synchronising MIDI sequences to recorded music, or fitting music which has been recorded 'rubato' to bar intervals.

6.3 Real-time synchronisation

There are a number of ways of organising real-time synchronisation in a MIDI system, but they all depend on the use of timecode in one form or another. In this section the principles of time code and its relationship to MIDI are explained, whilst practical examples of timecode synchronisation are covered in section 7.6.5.

6.3.1 What is timecode?

Timecode is more correctly referred to as SMPTE/EBU time and control code. It is often just referred to as SMPTE ('simpty') in studios. It comes in two forms: linear timecode (LTC), which is an audio signal capable of being recorded on a tape recorder, and vertical interval timecode (VITC), which is recorded in the vertical interval of a television picture. Timecode is basically a binary data signal registering time from an arbitrary start point (which may be the time of day) in hours, minutes, seconds and frames, against which the programme runs. It was originally designed for video editing, and every single frame on a particular video tape has its own unique number called the timecode address. This can be used to pinpoint a precise editing position. More recently timecode has found its way into audio, where TV frames have less meaning but are still used as a convenient subdivision of a second.

A number of frame rates are available, depending on the television standard to which they relate, the frame rate being the number of still frames per second used to give the impression of continuous motion in the TV picture. 30 frames per second (fps), or true SMPTE, was used for monochrome American television; 29.97 fps is used for colour NTSC television (mainly USA, Japan and parts of the Middle East), and is called 'SMPTE drop-frame'; 25 fps is used for PAL and SECAM TV and is called 'EBU' (Europe, Australia, etc.); and 24 fps is used for some film work. SMPTE drop frame timecode is so called because in order to maintain sync with NTSC colour television pictures running at 29.97 fps it is necessary to use the 30 fps SMPTE code but to drop two frames at the start of each minute, except every tenth minute. This is a compromise solution which has the effect of introducing a short term sync error between timecode and real time, whilst maintaining reasonable control over the long-term drift.

An LTC frame value is represented by an 80 bit binary 'word', split principally into groups of 4 bits, with each 4 bits representing a particular parameter such as tens of hours, units of hours, and so forth, in BCD (binary-coded decimal) form (see Figure 6.1). Sometimes, not all four bits per group are required – the hours only go up to '23', for example – and in these cases the remaining bits are either used for special control purposes or set to zero (unassigned): 26 bits in total are used for time address information to give each frame its unique hours, minutes, seconds, frame value; 32 are 'user bits' and can be used for encoding information such as reel number, scene number, day of the month and the like; bit 10 denotes drop-frame mode if a binary 1 is encoded there, and bit 11 can denote colour frame mode if a binary 1 is encoded (used in video editing). The end of each word consists of 16 bits in a unique sequence, called the 'sync word', and this is used to mark the boundary between one frame and the next. It also allows a timecode reader to tell in which direction the code is being read, since the sync word begins with '11' in one direction and '10' in the other.

This binary information cannot be recorded to tape directly, since its bandwidth would be too wide, so it is modulated in a simple scheme known as 'biphase mark', or FM, in which a transition from one state to the other (low to high or high to low) occurs at the edge of each bit period, but an additional transition is forced within the period to denote a binary 1 (see Figure 6.2). The result looks rather like a square wave with two frequencies, depending on the presence of ones and zeros in the code. The code can be read forwards or backwards, and phase inverted. Readers are available which will read timecode over a very wide range of speeds, from around 0.1 to 200 times play speed. The rise-time of the signal, that is the time it takes to

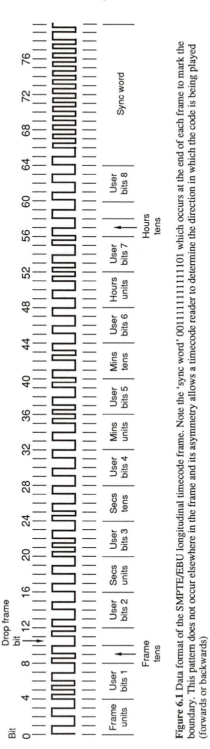

Figure 6.1 Data format of the SMPTE/EBU longitudinal timecode frame. Note the 'sync word' 0011111111111101 which occurs at the end of each frame to mark the boundary. This pattern does not occur elsewhere in the frame and its asymmetry allows a timecode reader to determine the direction in which the code is being played (forwards or backwards)

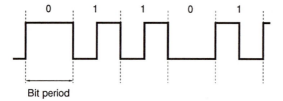

Figure 6.2 The FM or biphase-mark channel code is used to modulate the timecode data so that it can be recorded as an audio signal

swing between its two extremes, is specified as 25 μs ± 5 μs, and this requires an audio bandwidth of about 10 kHz.

VITC is recorded not on an audio track, but in the vertical sync period of a video picture, such that it can always be read when video is capable of being read, such as in slow-motion and pause modes. It is thus very useful in applications where slow-motion cueing is to be used in the location of sync or edit points. It is extracted directly from the video signal by a timecode reader. Some MIDI synchronisers can accept VITC, but this is much less common than the ability to read and write LTC.

6.3.2 Recording timecode

Timecode may be recorded or 'striped' on to tape before, during or after the programme material is recorded, depending on the application. Normally the timecode must be locked to the same speed reference as that used to lock the speed of the tape machine, otherwise a long-term drift can build up between the passage of time on the tape and the measured passage in terms of timecode. When working with video, such a reference is usually provided in the form of a video composite sync signal, and video sync inputs are increasingly provided on digital audio recorders for this purpose. In digital audio systems the timecode should also be locked to the audio sampling rate.

Timecode generators are available in a number of forms, either as stand-alone devices, as part of a synchroniser or editor, or integrally within a tape recorder. In large centres timecode is sometimes centrally distributed and available on a jackfield point. When generated externally, timecode normally appears as an audio signal on an XLR connector or jack, and this should be routed to the track required for timecode on the tape recorder, or to the timecode input of a disk-based digital recorder. Most generators allow the user to preset the start time and the frame rate standard. In MIDI systems, timecode is sometimes generated by the intelligent MIDI interface attached to a computer, which also acts to receive timecode information and convert it into the MIDI TimeCode (MTC) format. In such cases the frame rate and start time are usually selected within the software applications themselves.

Timecode is often recorded on to an outside track of a multitrack tape machine (usually track 24), or a separate timecode or cue track may be provided on digital machines. In disk-based digital recording systems timecode is used to reference recording and replay, but is not physically recorded as an audio signal. On tape machines the timecode signal is recorded at around 10 dB below reference level, and crosstalk between tracks or cables is often a problem due to the very audible mid-frequency nature of timecode. Some quarter-inch analogue machines have a facility for recording timecode in a track which runs down the centre of the guard band. This is called 'centre-track timecode'. Professional DAT machines are often capable of recording timecode, this being converted internally into a DAT running-time code which is recorded in the subcode area of the digital recording. On replay, any frame rate of timecode can be replayed using a conversion process, no matter what was used during recording, which is useful in mixed standard environments.

Timecode should run for at least 20 seconds or more before the recorded material begins in order to give other machines and computers time to lock in, and it should normally run contiguously (that is with the numbers incrementing upwards in an unbroken sequence) through the recording. In MIDI systems, the frame rate chosen should normally be that of the television standard used in the country concerned, unless there is a strong reason to do otherwise.

6.3.3 Principles of MIDI timecode (MTC)

MIDI timecode has two specific functions. Firstly, to provide a means for distributing conventional SMPTE/EBU timecode data around a MIDI system in a format which is compatible with the MIDI protocol, and secondly to provide a means for transmitting 'setup' messages which may be downloaded from a controlling computer to receivers in order to program them with cue points at which certain events are to take place. The intention is that receivers will then read incoming MTC as the program proceeds, executing the pre-programmed events defined in the setup messages. Sequencers and some digital audio systems often use MIDI timecode derived from an external synchroniser or MIDI peripheral when locking to video or to another sequencer. MTC is an alternative to MIDI clocks and song pointers, for use when real time synchronisation is important.

In an LTC timecode frame, two binary data groups are allocated to each of hours, minutes, seconds and frames, these groups representing the tens and units of each, so there are eight binary groups in total representing the time value of a frame. In order to transmit this information over MIDI, it has to be turned into a format which is compatible with other MIDI data (i.e. a status byte followed by relevant data bytes). There are two types of MTC synchronising message: one which updates a receiver regularly with running timecode, and another which transmits one-time updates of the timecode position for situations such as exist during the high speed spooling of tape machines, where regular updating of each single frame would involve too great a rate of transmitted data. The former is known as a quarter-frame message, denoted by the status byte (&F1), whilst the latter is known as a full-frame message and is transmitted as a universal realtime SysEx message.

6.3.4 MTC quarter-frame message

One timecode frame is represented by too much information to be sent in one standard MIDI message, so it is broken down into eight separate messages. Each message of the group of eight represents a part of the timecode frame value, as shown in Figure 6.3, and takes the general form:

&[F1] [DATA]

The data byte begins with zero (as always), and the next seven bits of the data word are made up of a 3 bit code defining whether the message represents hours, minutes, seconds or frames, MSnibble or LSnibble, followed by the four bits representing the

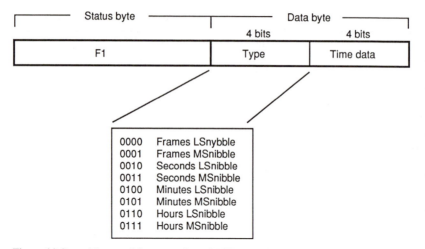

Figure 6.3 General format of the quarter-frame MTC message

binary value of that nibble. In order to reassemble the correct timecode value from the eight quarter-frame messages, the LS and MS nibbles of hours, minutes, seconds and frames are each paired within the receiver to form 8 bit words as follows:

Frames: rrr qqqqq

where 'rrr' is reserved for future use and 'qqqqq' represents the frames value from 0 to 29;

Seconds: rr qqqqqq

where 'rr' is reserved for future use and 'qqqqqq' represents the seconds value from 0 to 59;

Minutes: rr qqqqqq

as for seconds; and

Hours: r qq ppppp

where 'r' is undefined, 'qq' represents the timecode type, and 'ppppp' is the hours value from 0 to 23. The timecode frame rate is denoted as follows in the 'qq' part of the hours value: 00 = 24 fps; 01 = 25 fps; 10 = 30 fps drop-frame; 11 = 30 fps non-drop-frame. Unassigned bits should be set to zero.

At a frame rate of 30 fps, quarter-frame messages would be sent over MIDI at a rate of 120 messages per second. As eight messages are needed fully to represent a frame, it can be appreciated that $30 \times 8 = 240$ messages really ought to be transmitted per second if the receiving device were to be updated every frame, but this would involve what has been considered to be too great an overhead in transmitted data, therefore the receiving device is updated every two frames. If MTC is transmitted continuously over MIDI it will take up approximately 7.5% of the available data bandwidth. Quarter-frame messages may be transmitted in forward or reverse order, to emulate timecode running either forwards or backwards, with the 'frames LSnibble' message transmitted on the frame boundary of the timecode frame that it represents.

The receiver must in fact maintain a two-frame offset between displayed timecode and received timecode since the frame value has taken two frames to transmit completely. For real-time synchronisation purposes, the receiver may wish simply to note that time has advanced another quarter of a frame at the receipt of each quarter-frame message, rather as it advances by one-sixth of a beat on receipt of each MIDI clock. Internal synchronisation software should normally be able to flywheel or interpolate between received synchronisation messages in order to obtain higher internal resolution than that implied by the rate of the messages. For all except the fastest musical tempo values, MIDI timecode messages arrive more regularly than MIDI clocks would, and thus they might be considered a more reliable timing reference. Nonetheless, MIDI clocks are still needed when synchronisation is required to be based on musical time increments.

6.3.5 Full-frame message

As well as continuous update of timecode location over MIDI, the format allows for one-time general updates using a message which transmits a whole timecode frame value in ten bytes at one go. This can be useful when running timecode at high speeds during fast winding of tape machines, sending full-frame messages at intervals so that the receiving devices are kept in step with the transmitter, and this may avoid overloading the bus. The format of this message is as follows, falling into the group of messages known as the universal realtime messages (see section 2.5.13):

& [F0] [7F] [dev. ID] [01] [01] [hh] [mm] [ss] [fr] [F7]

The device ID would normally be set to &7F which signifies that the message is intended for the whole system, the sub-ID #1 of &01 denotes an MTC message, and sub-ID #2 denotes a full-frame message. Thereafter hours, minutes, seconds and frames take the same form as for quarter-frame messages.

6.3.6 User bits

The user bits of the SMPTE/EBU timecode frame may also be transmitted over MIDI using another system exclusive message (fifteen bytes). The user bits often contain static information in ASCII form which are often used for labelling tapes with origination information, but which may be put to any purpose, including the representation of a second timecode value. The general format of the user bit message is:

&[F0] [7F] [dev. ID] [01] [02] [U1] [U2] [U3] [U4] [U5] [U6] [U8] [U9] [F7]

The first four bits of each of U1 to U8 are set to zero, whilst the latter four bits of each correspond to the eight binary groups of user bits from the timecode frame. U9 contains six zeros in the most significant positions, and its two LSBs represent the two binary group flag bits from the timecode frame.

6.3.7 MTC cueing and setup messages

The setup messages transmit information about the locations of cue points, record drop-ins, event starts, and so on, to peripheral devices. A summary will be given in the following section.

By means of a setup message, a controlling device may program a receiver so that it is ready to execute certain functions at specific timecode locations. When the receiver decodes the relevant MTC quarter-frame value it will normally execute the pre-programmed function. The general format of a setup message is as follows, being a universal non-realtime message (see section 2.5.13):

& [F0] [7E] [dev. ID] [04] [sub ID 2] [hh] [mm] [ss] [fr] [ff] [sl] [sm] [info] ...
... [F7]

The sub-ID #1 of &04 is to identify the message as an MTC message; sub ID #2 is used to denote the type of setup message (see Table 6.2); [hh mm ss fr ff] indicate the time at which the event is to occur (in the same format as other MTC messages, except including an extra byte [ff] to denote fractions of frames, or one-hundredths of frames); [sl] and [sm] denote the event number, transmitted LSB first; and 'info' provides a means for including additional data with the message. Additional data, which is often MIDI event data to occur at the particular time defined in the message, is transmitted in a 'nibblised' form, with the least significant nibble of the MIDI message transmitted first. In nibblised form, each MIDI message byte is split into two nibbles, and transmitted with leading zeros. For example, the status byte &B3 would be transmitted as &03 &0B.

The 'special' group of events are different to normal events. When sub-ID #2 is set to 00 the event number becomes a definition of the type of special message, as shown in Table 6.3.

Table 6.2 MTC setup identifications

Sub-ID #2	Function
&00	Special
&01	Punch-in
&02	Punch-out
&03	Delete punch-in
&04	Delete punch-out
&05	Event start
&06	Event stop
&07	Event start plus additional info.
&08	Event stop plus additional info.
&09	Delete event start
&0A	Delete event stop
&0B	Cue point
&0C	Cue point with additional info.
&0D	Delete cue point
&0E	Event name in additional info.

Table 6.3 Special functions (affect entire device)

[sl] [sm]	Function
00 00	Timecode offset
01 00	Enable event list
02 00	Disable event list
03 00	Clear event list
04 00	System stop
05 00	Event list request

There is also a group of messages almost identical to the setup messages but intended for real-time use. These would be used when an event was to be triggered remotely in real time, rather than being programmed to occur at a forthcoming MTC time. Since the application is real time, the message takes the universal realtime header:

&[F0] [7F] [dev. ID] [05] [sub-ID #2] [sl] [sm] [info] ... ... [F7]

Sub-ID #2 is identical to the non-realtime version as shown in Table 6.2, except that the delete messages do not exist (they do not need to, because the event is not being stored), and only the 'system stop' special message is used. The event time is not included because this is a real-time trigger message.

6.3.8 Direct time lock (DTL)

Direct time lock is not part of the MIDI standard, but some MIDI hardware and software uses it as an alternative to MTC for real-time synchronisation to a timecode track. It was developed by Southworth Music Systems and subsequently used by Mark of the Unicorn in its widely used MIDI TimePiece (a multiport MIDI interface with timecode facilities for the Macintosh).

DTL uses a system exclusive message with the manufacturer ID of &28 to transmit its equivalent of a full-frame MTC message, followed by frame advance messages which (rather unusually) use the same status byte as MIDI clocks (&F8). Every so often new full-frame messages are sent which update the receiver. DTLe (Enhanced DTL) was introduced more recently, and is similar to MTC in that four frame advance messages are sent per frame as opposed to the one of DTL.

6.4 Remote machine control

6.4.1 Overview of MIDI machine control (MMC)

MIDI may be used for remotely controlling tape machines and other studio equipment, as well as musical instruments. The MTC cueing messages introduced above are a first taste of the possibilities available here, but there is a further massive set of MIDI messages designed specifically for what is known as MIDI Machine Control or MMC. MMC uses universal realtime SysEx messages with a sub-ID #1 of either &06 or &07, and has a lot in common with a remote control protocol known as 'ESbus' which was devised by the EBU and SMPTE as a universal standard for the remote control of tape machines, VTRs and other studio equipment. The ESbus standard uses an RS422 remote control bus running at 38.4 kbaud, whereas the MMC standard uses the MIDI bus for similar commands. Although MMC and ESbus are not the same, and the message protocols are not identical, the command types and reporting capabilities required of machines are very similar.

As with much about MIDI, the MMC specification is worded in a looser fashion than the ESbus specification, and appears to be designed to give manufacturers rather greater flexibility in implementation. More is left to chance in reality. There are a number of levels of complexity at which MMC can be made to operate, and communication is possible in both open and closed loop modes. By allowing this flexibility, MMC makes it possible for people to implement it at anything from a very simple level (i.e. cheaply) to a very complicated level involving all the finer points. MMC is gaining increasing popularity in semi-professional equipment, because it is somewhat cheaper to implement than ESbus and allows equipment to be integrated easily with a MIDI-based studio system. There are a number of tape machines and synchronisers now on the market with MIDI interfaces, and some sequencer packages handle the remote control of studio machines using the MMC protocol. Studio machines may be connected to the main studio computer by connecting them to one port on a multiport MIDI interface. Some examples are given in Chapter 7.

6.4.2 Open and closed loops

As mentioned above, MMC is designed to work in either open or closed-loop modes (see Figure 6.4), and uses system exclusive messages to control remote machines. This is similar to other system exclusive applications which can make use of handshaking between the transmitter and the receiver, such as sample dumps and file transfers. Communication can be considered as occurring between a 'controller' and a 'controlled device', with commands flowing from the controller to the controlled device and responses returning in the opposite direction. Since a controller may address more than one controlled device at a time it is possible for a number of responses to be returned, and this situation requires careful handling, as discussed below.

It is expected that MMC devices and applications will default to the closed-loop condition, but a controller should be able to detect an open-loop situation by timing out if it does not receive a response within two seconds after it has sent a message which requires one. From then on, an open loop should be assumed. Alternatively, a controller could continue to check for the completion of a closed loop by sending out regular requests for a response, changing modes after a response was received. In the closed-loop mode a simple handshaking protocol is defined, again similar in concept to the sample and file dump modes, but involving only two messages – WAIT and RESUME. These handshaking messages are used to control the flow of data between controller and controlled device in both directions, in order to prevent the overflowing of MIDI receive buffers (which would result in loss of data). Handshaking is discussed further below.

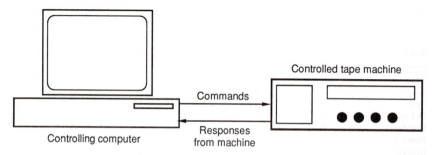

Figure 6.4 A closed-loop MMC arrangement. The controller should receive a response from the controlled device within two seconds of issuing a command which expects a response. If it does not, it should assume an open loop

6.4.3 MMC commands

Typical MMC communications involve the transmission of a command from the controller to a particular device, using its device ID as a means of identifying the destination of the command. It is also possible to address all controlled devices on the bus using the &7F device ID in place of the individual ID. Commands take the general format:

&[F0] [7F] [dev. ID] [06] [data] ... ... [F7]

Note that only sub-ID #1 is used here, following the device ID, and there is no sub-ID #2 in order to conserve data bandwidth. The sub-ID #1 of &06 denotes an MMC command. [data] represents the data messages forming the command, and may be from one to many bytes in length.

The amount of data making up a command depends on its type. Commands which consist of only a single byte, such as the 'play' command (&02), occupy the range from &01 to &3F (&00 is reserved to be used for future extensions to the command set). A typical command of this type (e.g. 'play') would thus be transmitted as:

&[F0] [7F] [dev. ID] [06] [02] [7F]

Table 6.4 gives a list of the single byte transport commands used in the MMC protocol.

Longer commands have a number of data bytes attached to them, and occupy the range from &40 to &77. In order that a receiver knows how many bytes to expect in the command message, a 'count' byte follows the command identifier, indicating how many data bytes follow. For example, the 'variable play' command (&45) which instructs a tape machine to play in variable speed mode includes three data bytes to indicate the speed and direction of play, and includes a count byte of &03:

&[F0] [7F] [dev. ID] [06] [45] [03] [data] [data] [data] [F7]

Messages where the number of data bytes following the sub-ID #1 are greater than the allowed MMC size of 48 bytes can be segmented, by spreading them across a number of SysEx messages using a segmentation technique described in the standard.

The particular commands 'write' (&40) and 'read' (&42) are used to address the internal registers of a controlled device, in order to perform such operations as setting a timecode generator's start time. The 'read' commands requires a response from the controlled device, as described below. Since the list of MMC commands

Table 6.4 Basic MMC transport controls

Command	Hex value	Comment
Stop	01	
Play	02	
Deferred play	03	Play after autolocate achieved
Fast fwd	04	
Rewind	05	
Record strobe	06	Drop into or out of record (depending on rec. ready state)
Record exit	07	
Record pause	08	Enters record-pause mode
Pause	09	

and their accompanying data occupies many pages it is not proposed to describe them in detail here. Readers should refer to the MIDI Machine Control documentation for the latest information (see Appendix 2).

6.4.4 MMC responses

Certain MMC commands, such as 'read', require responses from the controlled device, and these are returned to the controller using the sub-ID #1 of &07 (as opposed to the command sub-ID of &06). Long responses (longer than 48 bytes) may be segmented in the same way as commands.

Responses may have a number of data bytes following them, depending on the type of response. Certain ranges of response identifiers are set aside for particular lengths of data, in order that the controller knows what to expect. Where the data length is variable, a 'count' byte is included, as for commands. Responses in the range &01 to &1F always have 5 data bytes and are used for the transmission of time code values; the range &20 to &3F always has two data bytes and is used for the transmission of so-called 'short' timecode values (where only the frames and subframes are updated); &40 to &77 are variable length responses using a count byte after the response type to define the length. A response therefore takes the general form:

&[F0] [7F] [dev. ID] [07] [response type] [data] … … [7F]

Responses are used for such purposes as returning timecode values to a controller from a tape machine's internal timecode reader, or for reporting on the current status of the machine (such as which tracks are currently record enabled). The response returns a value stored in one of the tape machine's internal registers, and these registers will be updated by the tape machine's internal CPU. Error handling techniques are described in the standard, to deal with situations where a device is unable to provide the requested information, or where it is unable to respond to a particular command.

6.4.5 MMC handshaking

The handshaking messages, WAIT (&7C) and RESUME (&7F), can be issued by either the controller or any of its controlled devices. Handshaking depends on the use of a closed loop.

When issued by the controller the message would normally be a command addressed to any device trying to send data back to it, and thus the device ID attached to controller handshaking messages is &7F ('all call'). For example, a controller whose receive data buffer was approaching overflow would wish to send out a general 'everybody WAIT' command, to suspend MMC transmission from controlled devices until it had reduced the contents of the buffer, after which an 'everybody RESUME' command would be transmitted. Such a command would take the form:

&[F0] [7F] [7F] [06] [7C or 7F] [F7]

When issued by a controlled device, handshaking messages should be a response tagged with the device's own ID, as a means of indicating to the controller *which* device is requesting a WAIT or RESUME. On receipt of a WAIT from a particular device ID the controller would suspend transmissions to that device but continue to transmit commands to others. Such a message would take the form:

&[F0] [7F] [dev. ID] [07] [7C or 7F] [F7]

6.4.6 Implementing MMC

MMC could be used simply to control the basic transport functions of an audio tape recorder. In such a case only a very limited set of commands would need to be implemented in the tape recorder, and very little would be needed in the way of responses. In fact it would be quite feasible to operate the transport of a tape recorder using an open-loop approach – simply sending 'play', 'stop', 'rewind', etc. as required by the controlling application. In a system where synchronisation and timecode generation was important there would be a greater need for a closed loop and more reporting capabilities on behalf of the tape recorder, and some system examples of this type are shown in section 7.6.6.

There is no mandatory set of commands or responses defined in the standard, although there are some guidelines concerning possible minimum sets for certain applications. It is possible to tell which MMC commands and responses have been implemented in a particular device by analysing the 'signature' of the device. The signature will normally be both published in written form in the manual, and available as a response from the controlled device. It exists in the form of a bit map in which each bit corresponds to a certain MMC function. If the bit is set to '1' then the function is implemented. The signature comes in two parts: the first describing the commands implemented and the second describing the responses implemented. It also contains a header describing the version of MMC used in the device. The exact format of the signature is described in the MMC standard.

6.5 Synchronising software applications

6.5.1 Synchronising a sequencer to an external sync source

To run a software application such as a sequencer locked to an external source of timing data such as MIDI clocks or MTC it is necessary to switch it to external sync mode. Some sequencers will also allow automatic selection of the sync source, in order that replay will follow external data if it is present but internal clock if not. It is normally necessary also to define which type of sync data is expected. When in an external sync mode, some sequencers need to be put into a 'play ready' mode in order that they will start to play when sync information is received, and this is often done by pressing the play button, upon which the button flashes, waiting for external commands before actually playing.

6.5.2 Synchronising multiple applications to a common clock

So far the topic of synchronisation and machine control has been covered from an 'external' MIDI point of view – in other words, when devices are connected together using MIDI interfaces – but there is also a growing need to synchronise multiple software applications which run on the same computer. This is achieved in different ways depending on the operating system and hardware in use, but most systems now offer some form of multitasking whereby it is possible to run more than one application at a time. For example, it might be desired to run a MIDI mixer automation system in parallel with a MIDI sequencer, with them both locked to the same source of timing information. Or it might be desired to use a development application such as MAX (see section 5.10) as the master source of timing data for a sequencer.

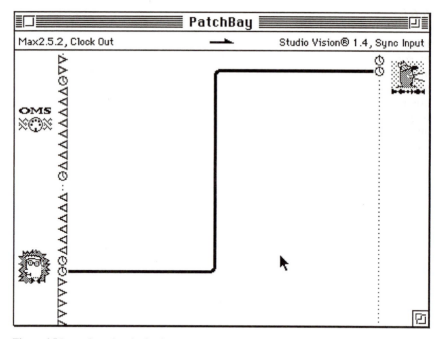

Figure 6.5 Internal synchronisation between two MIDI applications running on a computer using Apple's MIDI Manager. The clock icons of the two applications concerned are linked

A fast computer is needed to work properly with multitasking MIDI applications, and even with a fast machine there can be considerable problems with synchronisation 'jitter' since the processor is sharing its time between the different open applications. It is also important to ensure that the applications you intend to use actually support co-timed operation, since although they may be able to run simultaneously, only one of them at a time may be able to synchronise to the clock source. Recent versions of software may support the equivalent of 'inter-application communications', which allows information from one software application to be used by another, and this has been used by some sequencer manufacturers to set up a 'hot link' between two applications for the purpose of synchronisation.

Apple Macintosh users will also be familiar with MIDI Manager which for a long time was the only way to route timing data between applications and allow them to share access to timing data. The example in Figure 6.5 shows a MIDI Manager patch used to connect the clock icons of two applications together. Here Opcode's MAX is used as a master sync source to lock the replay of Vision, which is set to work in external sync mode. MIDI Manager also acts as a form of synchronisation gearbox, converting timing information in one form to another (say from beat clock to timecode). It is likely that future versions of Macintosh MIDI operating systems such as OMS will manage the synchronisation of multiple applications without the need for MIDI Manager, using inter-application communications.

6.6 MIDI batons

It is important to mention here the growing interest in products which allow a user to 'conduct' an electronically stored score. Most of these products are experimental and exist in various forms of development, and using various means of detection generate synchronisation data which can be used to control the replay tempo and possibly other expressive parameters of a sequenced musical score. At the present time these are relatively crude devices which respond to simple beat patterns or to an up–down beat, but further work is likely to result in greater development of such batons which may allow them to be used more freely.

Chapter 7

Practical system design

This chapter looks at MIDI systems issues. It describes examples of practical systems for different applications, and includes information about audio routing and mixing, as well as showing how a MIDI system may be synchronised with other studio equipment.

7.1 Introduction to systems

Simple interconnection of MIDI equipment was introduced in Chapter 2, showing how one device transmits to a number of receivers. Once the system becomes more complicated, though, such as with the addition of effects units and additional sound generators, the routing requirements for MIDI data may need more careful thought, especially if the constant replugging of cables is to be avoided. If, for example, the computer is to be used for sequencing as well as for voice or sample dumps, then a means will be required by which the MIDI input of the computer may be fed from a variety of possible sources, depending on which is dumping data or acting as the master keyboard. For many of the purposes described in the foregoing chapters, a two-way link is often required between the MIDI device concerned and the computer, in order that ordered dumps of data with handshaking can be executed. The user must therefore think about incorporating routers for MIDI data, which allow the connection of certain sources to certain destinations, perhaps with some processing intervention (see below).

A further consideration will be that of how to handle audio signals. The addition of a mixer, perhaps automated using MIDI (see Chapter 4) or using its own automation system locked to timecode, will become almost mandatory, as will the possible addition of a MIDI-controlled audio routing matrix which takes in a number of sources and allows each input to be fed to a chosen output or outputs.

Although a considerable amount of multitrack work may be carried out with a 'MIDI-only' system, there is likely to come a point when some form of audio recording becomes necessary, in order to record 'real' tracks of audio, as opposed to 'virtual' MIDI tracks, and for stereo recording of mixed audio signals. This might take the form of a multitrack tape recorder (either analogue or digital) or perhaps a digital disk-based recording system integrated with the sequencer. In the case of an external tape recorder a means will be required of synchronising the MIDI system to it or vice versa, so that the replay of sequenced sound may be made to track the replay of real sounds. Using MIDI Machine Control and MIDI Show Control it may also be possible to control external equipment remotely.

When MIDI installations cover a number of studio areas which are situated apart from each other, or where the number of connections and cable lengths to be covered exceed the capabilities of standard MIDI interfaces and cables, it is possible to incorporate a high speed network into the system design. As discussed in section 7.8 it is possible for a data network, using either fibre optic or copper connections, to carry packets of control data for all sorts of applications including MIDI, and to route these packets to one of a number of 'gateways' attached to the network. The number of sources and destinations can then be expanded considerably and interconnections can be made over much longer distances than normal, particularly where fibre optics are concerned.

7.2 The master keyboard and expanders concept

All MIDI systems must be controlled by a source of MIDI data, which will feed the inputs to some or all of the devices in the system. This source will be referred to as the controller, whatever it happens to be in physical terms. Most commonly, in a recording or music composition environment, the controller will be a desktop computer, running one or more of a number of possible commercial MIDI software packages. In a performance environment it is possible that the controller might be a musical keyboard, but many performers also make use of pre-sequenced tracks for backing.

Even if the overall system controller is a computer, there will nearly always be the need for a master keyboard, since the computer must originally derive its data from a MIDI source. It is possible that the master keyboard could simply be one of the sound-producing keyboards within the greater system, in which case it will also receive data from the computer as well as providing the computer with a source of musical and control information. This keyboard should be a comprehensive piano-type keyboard with all of the facilities required to control other instruments, since this will be the most often used source of MIDI information for the rest of the system. A number of considerations arise from this suggestion: firstly, the master keyboard must be able to send all the different controller messages that are commonly required, such as velocity, after-touch, pitch bend, modulation wheel, pedals, program changes and perhaps messages from other analogue sliders or wheels; secondly, the master keyboard should have a full-range keyboard to make maximum use of the receivers' sound generators; thirdly, one must remember that the master keyboard's own audio output may sound when its keyboard is played, even though it is transmitting to another device, so either its output level would have to be cut, or a means of implementing the local off function would be required; and finally, the other instruments in the system need not have keyboards of their own, as they could be 'played' under MIDI control from the master.

Some manufacturers have taken up the idea of making a master keyboard to fulfil these functions, which is not intended to produce sound but has a high quality keyboard and a wide range of control features, but the approach has not proved to be particularly popular. The lack of popularity may be attributed to the relatively high cost of some master keyboards, coupled with the fact that people have not taken to the idea of spending a lot of money for a keyboard which doesn't make sounds of its own. Consequently, the master keyboard in many systems is often a good synthesiser.

The concept of the expander, that is a device which can generate sounds under MIDI control but which is without a keyboard of its own, has proved popular in systems where a large number of sounds are required, and where the controller is a sequencer.

Many manufacturers make expanders as versions of their more popular keyboards, perhaps with some variations in number of outputs or facilities, which can be rack mounted. They contain all the MIDI and voice generation circuitry in a fraction of the space taken up by the corresponding keyboard version.

7.3 Basic MIDI device interconnection

In Chapter 2 the simple daisy-chain interconnection of MIDI devices was introduced, but this is only really satisfactory for small systems and there are other methods of interconnection which will often be more appropriate and may result in more reliable operation, especially where larger systems are concerned.

In a MIDI system devices needing to receive MIDI data may either be individually connected to a central distribution point, or, alternatively, one bus may be passed from device to device by looping through (the so-called 'daisy chain'). The former is rather like the older method of domestic electrical wiring known as the 'spur' system, in which all sockets are wired back on separate spurs to the mains supply, and the latter is a little bit like the domestic ring main, where one 'ring' goes round the whole house and each socket is taken off the ring (except that the ring is not completed in the MIDI system). It is also possible to use a combination of the two methods. The diagrams in Figure 7.1 show some different possibilities for a system in which one controlling device (pictured as a computer) is set up to replay prerecorded data to a number of receivers.

It can be seen in Figure 7.1(a) that the MIDI THRU is used to pass on the data from the controller to the next device, but that the system in Figure 7.1(b) makes no use of the MIDI THRU on each instrument, as each is fed from a central collection of MIDI THRUs housed in a distribution box. In its simplest form the distribution box is often known as a 'MIDI THRU' unit, as it provides a number of THRUs from one IN, and it means that the signal arriving at each device in the system will have only passed via one THRU stage and perhaps a limited length of cable, rather than the large number of THRUs in the previous daisy chain. It is advantageous for MIDI data to have passed through as few THRU links as possible because each stage adds distortion to the data signal which could eventually result in receive errors at the end of the chain. (In the simple systems illustrated all the MIDI data from the controlling device will be fed to every instrument in the system.)

A MIDI THRU box having two outputs fed from one input could be used in cases when a device does not have its own THRU port, and when the user needs to daisy-chain the bus on to another device. One THRU could feed the first device (the one without the THRU port) and the other could feed the next device in the chain. The alternative is to put a device without a THRU at the end of a daisy chain.

7.4 Delays in MIDI systems

Delays are a subject of considerable misunderstanding amongst people that use MIDI, and a number of myths have arisen over the years. Most of the time most people using MIDI will not notice a problem with delay, but it is worth adopting good practice in system design in order to minimise log-jams in the system which might result in buffer overflows or delays in certain operational cases. A correct approach will also help to improve reliability and overall system timing integrity.

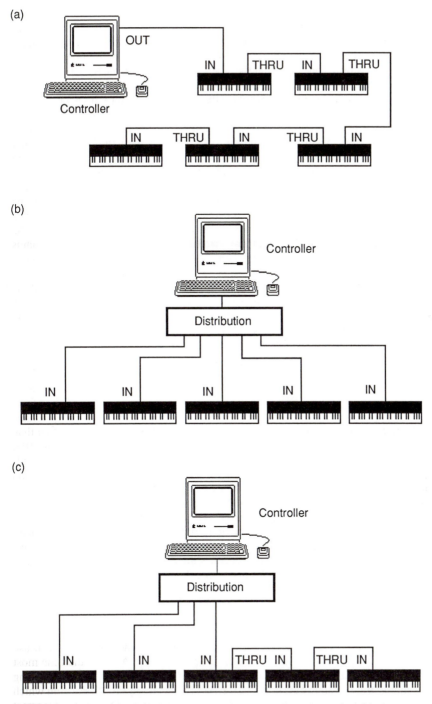

Figure 7.1 Examples of simple interconnection between a controller and a number of devices.
(a) THRU chain; (b) using a distribution matrix; (c) using a combination of the two approaches

7.4.1 Serial delay

As MIDI is inherently a serial bus, all data must be sent sequentially, that is one data byte after another. A 6 note chord in which all the notes had been played together would be transmitted one note after the other. A 3 byte note on message takes nearly a millisecond to transmit, thus the 6 note chord would take 6 milliseconds, although running status could be invoked to reduce this by nearly one third. Intersperse these note messages with a sprinkling of real-time clock messages, some aftertouch information and a few controller messages and the delays can build up.

If a number of events are supposed to be transmitted at the same time by a sequencer to a number of devices then it is unavoidable that the events will not actually be transmitted together. The more data a sequencer is expected to transmit over one MIDI interface, the more difficult it will become to output messages for every channel at the correct time. Sequencer software and MIDI operating system extensions deal specifically with the difficult problem of ensuring that events are transmitted as close to their intended time as possible, and this demands prioritisation based on the urgency of the message. Clearly real-time messages will be given a higher priority than non-realtime, and controllers would normally be given a lower priority than note messages.

The data rate of MIDI is high enough to handle many real-time situations without perceptible delay. The human hearing mechanism begins to be able to perceive delays between sounds when they are greater than 20–50 ms apart, depending on the type of sound, and most real musicians do not play chords that are perfectly note-synchronous in any case. Apparent delays resulting from serial transmission, therefore, depend largely on the amount of data to be transmitted within a given period of time by the controller, since the baud rate of the interface is fixed.

7.4.2 Buffer delay

A further potential for delay arises because a device will not know if a message is intended for it without first examining the data, so devices must take time to look at every status byte that comes along, even though a large proportion of them may be intended for another destination. This process is called 'parsing'. It requires that devices have input buffers which can store incoming MIDI data until it can be dealt with, and the more devices share one bus the fuller these buffers will get. Some cheaper devices often display the 'MIDI BUFFER FULL' message when their limited memories fail them. A MIDI device will normally have a buffer for at least 128 bytes, but a larger buffer may not really be an advantage since this will only increase the potential for delay. Far better is the adoption of a faster CPU in the receiver, since this makes it possible to deal with data more rapidly, thus keeping the buffer empty. Some manufacturers now publish details of the time taken to produce an audio output after receiving a note on message. Currently the state of the art is about 2 ms.

7.4.3 Internal processing delays

MIDI receivers vary significantly as to their internal processing time, taking different amounts of time for sound generators to respond to incoming MIDI data, even disregarding any serial or buffer delays in the system. This may require the artificial delaying of MIDI data sent to 'faster' instruments in order to ensure a uniform overall delay.

It is really a myth that delays arise between the MIDI IN and the THRU of devices, as it was shown in Chapter 2 that there is little in between the IN and the THRU sockets to cause delays. Indeed, as the rise time of the opto-isolator in the hardware interface may not be longer than 2 μs, this is unlikely to introduce audible delay. If the slew rate of the opto-isolator or buffer amplifiers was slow enough to cause audible delay then MIDI would not work at all, as the decision level between a one and a zero would never be reached during the period of a bit cell, and thus no data would be registered. There is certainly no processing between these IN and THRU in a normal synthesiser or sampler, although the software THRU function of a sequencer may involve the data being read in and retransmitted from the OUT, possibly resulting in a delay of a few milliseconds.

7.5 Good system design

The best way to ensure that delays and buffer overflows do not become a problem is to minimise the amount of data transmitted to each device. Ideally a device should only be sent the data intended for it, using a dedicated cable from the controller, but the rule of thumb is not to send any more data than you have to down any one piece of wire. This may involve the use of (a) a controller with more than one MIDI interface, (b) the filtering out of unnecessary data before transmission, (c) the avoidance of THRU links to chains of devices, and (d) the use of external MIDI patchers or routers. Cable lengths in a MIDI system should be kept as short as possible. Long cables result in signal losses and distortion of the data, and this can result in errors in extreme cases. It is another myth that long cables result in delays.

The best approach to system design centred on a computer is to use a multiport MIDI interface attached via a high speed link to the computer, as described in Chapter 5. Here MIDI information is transferred between the interface and the computer at a higher rate than the normal 31.25 kbaud. The interface separates the messages based on its software configuration so that only certain messages are transmitted through certain ports. It may also be possible to filter out unwanted information such as timing data for some ports, if the device hanging on that port does not require it. Using such a method each MIDI interface can be entirely independent of the others, and, if the number of ports is large enough, each device can have its own port. When devices are capable of receiving in multi mode on all sixteen channels at once this can be a great benefit because it allows the number of MIDI channels in the system to be expanded by 16 times the number of ports.

A multiport approach can also be combined with any of the other approaches to MIDI interconnection, and this may be necessary if there are only a limited number of ports. An example is shown in Figure 7.2. Here a 4 port MIDI interface has three instruments connected to one of its ports using an external patcher. A patcher may have a number of inputs as well as outputs, allowing each of the three devices in the above example to be connected for both transmission and reception of MIDI data. The current configuration would be used to determine which inputs were routed to which outputs. A patcher is often able to respond to program change information received on a particular interface or channel and this can be used to recall stored routing configurations during the replay of a song if required.

A patcher or router may be able to investigate the data coming into it from a sequencer, determine what channel the messages are intended for, and route particular channel data to particular MIDI OUTs. This only transfers the channel

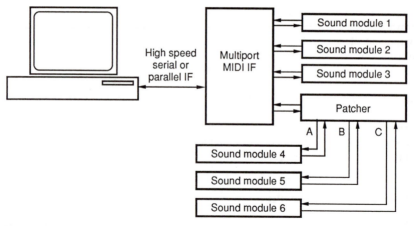

Figure 7.2 An external patcher can be used to manage the wiring of a larger system. The patcher in this example allows MIDI data from port 4 of the multiport MIDI interface to be routed to any or all of its output ports A, B and C. By programming the patcher appropriately, MIDI data from any of the three sound generators may be routed back to the computer via port 4 of the interface

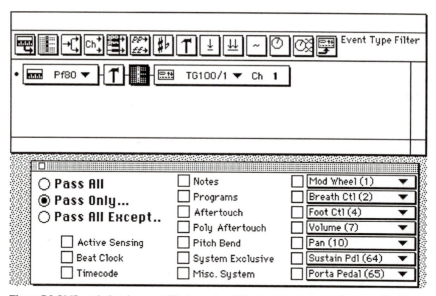

Figure 7.3 OMS patch showing event filtering and modification between a source and destination device. The hammer indicates a note velocity modification (scaling, limiting, etc.), and the following box indicates event filtering which is then specified in the window below

sorting job from the instruments to the routing matrix, and there is not really any reason why this should speed things up, but it may reduce the potential for buffer overload in the receivers. Some patchers have the capability to delay certain outputs by a given number of milliseconds in order to compensate for device-specific delays in other instruments, and they may also incorporate other MIDI processing features for creating virtual instruments as discussed in Chapter 5. The example in Figure 7.3 shows how MIDI message modification and filtering can be introduced between a controlling device and a receiver, in order that the receiver gets only certain information.

7.6 Some suggested system layouts

In this section some examples of MIDI systems will be given, ranging from the very simple to the advanced. Clearly these are not the only ways to configure a system, but they are designed to show approaches to many of the most common scenarios.

7.6.1 Basic music production system

Figure 7.4 shows a 'budget' configuration for a music production system, such as might be installed for basic educational purposes, for home use or for a first move into MIDI-based music production. The master keyboard also produces sound and would normally be used in the 'local off' mode in order that its sound generators and keyboard could be logically separated. The expander would be most versatile if it was capable of operation in 'multi' mode, able to receive information polyphonically on all 16 MIDI channels. A moderately-priced General MIDI sound module with built-in audio effects such as reverb would be suitable here. The 16 polyphonic voices could be level controlled and panned between left and right outputs internally using MIDI volume and pan controller data, obviating the need for an external audio mixer.

One possible configuration for the audio routing is shown here, taking advantage of the facility provided by some sound expanders to accept an external audio input

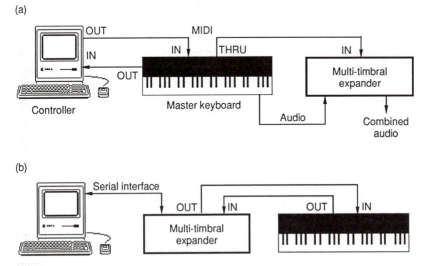

Figure 7.4 Example of an entry level computer-controlled MIDI system. (a) Conventional arrangement. (b) Using an expander with a built-in serial port for direct connection to the computer, with the expander acting as a MIDI interface for the computer

and mix it with their own output. A control is provided on the front panel of the expander to alter the mix level of the external source. The output of the expander could then be connected to a stereo audio recorder and monitor loudspeakers.

The computer would run any sequencer software and other librarians and editors required by the user, and would be interfaced to the MIDI equipment using a basic single port MIDI interface. Alternatively, since some sound expanders also act as single port MIDI interfaces, one could connect the expander directly to the computer using a standard serial interface cable.

7.6.2 Intermediate-level music production system

The system in the previous section could be expanded as needs increased. In Figure 7.5 is pictured an example of one possible configuration which adds a sampler, another sound module and a non-automated audio mixer. It is assumed that the sampler only has a stereo output, although samplers with multiple outputs could be accommodated by increasing the number of mixer channels.

A 4 port MIDI interface has been suggested here as a means of maintaining good order in the interconnections, connected using a fast serial port to the computer. Each port has both an IN and an OUT for bidirectional communications between the computer and the device, allowing voice data dumps to any of the sound generators or the sampler.

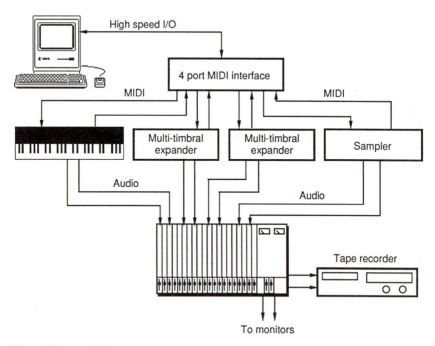

Figure 7.5 Example of an intermediate-level computer-controlled MIDI system for music production, showing stereo audio routing to a non-automated mixer. A stereo tape recorder can be used to store audio mixes

7.6.3 Advanced music production system

The system shown in Figure 7.6 is an example of a relatively advanced music production system. The computer is interfaced to the MIDI equipment using a multiport interface, and an automated mixer has been added, allowing dynamic control of the mix by the sequencer. Two audio effects units have also been added, and provided with MIDI control so that effects programs can be selected using program change messages and various other effects parameters adjusted under MIDI control (see Chapter 4). It would also be possible to incorporate a MIDI-controlled audio routing matrix such as described in section 4.4, if regular reconfigurations of the audio paths were anticipated.

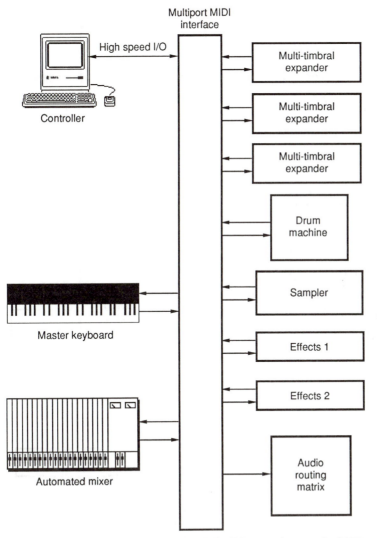

Figure 7.6 Example of a large computer-controlled MIDI system, incorporating MIDI-controlled audio effects and an automated mixer. The audio routing is not shown here for clarity

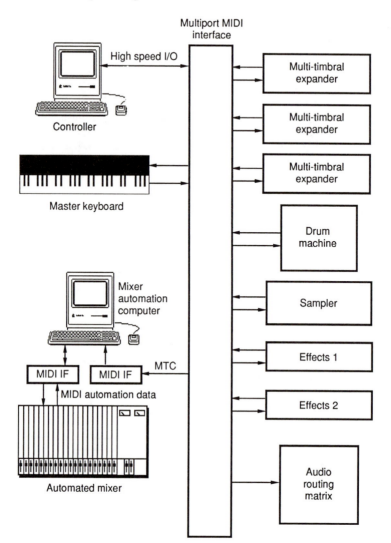

Figure 7.7 In this alternative configuration, an independent computer runs the mixer automation, synchronised to the sequencer running on the first computer using MTC routed to one of the ports on the multiport interface

A second option shown in Figure 7.7 uses a separate computer to handle the mixer automation, running a dedicated software package for this purpose rather than relying on the sequencer software to store mix data. (It would of course be possible as an alternative to run this software on the same computer as the sequencer, provided that suitable multitasking facilities were available.) Here the second computer is synchronised to the first using a MIDI sync signal (either beat clock or MTC) connected to a separate MIDI interface attached to the second serial port on the second computer.

7.6.4 Music notation system

The system shown in Figure 7.8 is designed for music publishing and notation. The capability for sound generation is not particularly high, but is provided to allow the composer or music copyist to replay the score either through headphones or loudspeakers as a means of checking the notated material. A laser printer has been added for the purpose of proofing the score. A scanner is optionally used for scanning in musical scores for the purpose of character recognition, so that music data may be quickly entered and stored on the computer. Specialised software is required for scanning, which will be able to produce files in one of the recognised musical file formats such as the MIDI file or DARMS notation, for importation into a music notation package or sequencer with scoring facilities.

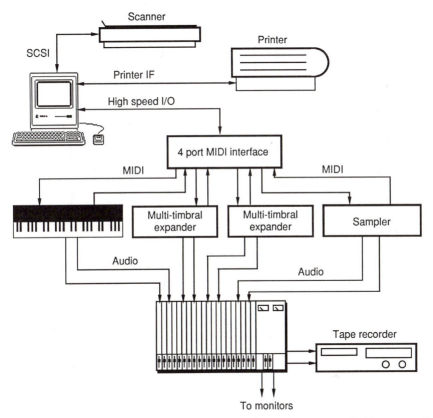

Figure 7.8 Example of a music notation system. A laser printer and scanner are added for output and input of printed scores, and a basic collection of sound sources are connected to the computer using MIDI, in order to replay a stored score. The MIDI keyboard is used during entry of a new score

7.6.5 External timecode synchronisation

Timecode synchronisation is used principally for locking the replay of a stored MIDI sequence to the replay of an external recorder. The external recorder might

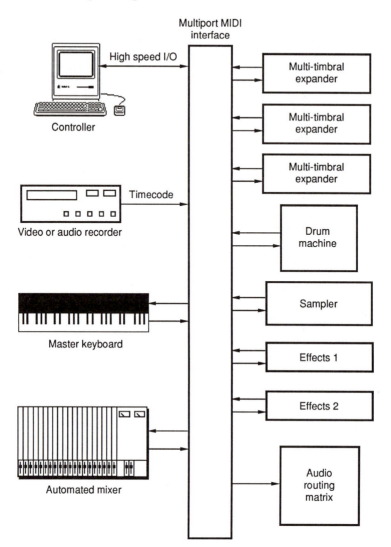

Figure 7.9 A video or audio recorder can be integrated into this system example, using replayed timecode to synchronise the replay of the MIDI sequencer running on the computer. This assumes that the multiport MIDI interface has a timecode input

be a multitrack audio recorder on which tracks of sound could be recorded in parallel with the MIDI-sequenced tracks in a studio. Alternatively it could be a video tape or disk recorder so that the replay of MIDI-sequenced material could be locked to the replay of video pictures, for the purpose of sound dubbing to picture (Figure 7.9).

In the illustrated examples timecode from the external machine is connected to the timecode input on a multiport MIDI interface. Such a timecode port is quite a common feature on the more advanced MIDI interfaces. A timecode reader in the MIDI interface normally converts the linear SMPTE/EBU timecode signal from the tape recorder into MIDI TimeCode (MTC) data (see Chapter 6), which allows the current time location of the tape recorder in hours, minutes, second and frames to be transmitted to the computer and any MIDI device which might require it. The software for the MIDI interface normally allows the user to determine which devices in the system will receive MTC data, and it is advisable not to send it to a device unless it is required because quarter-frame MTC messages consume a considerable proportion of the available MIDI bandwidth. Any variations in the speed of the timecode read by the interface (due perhaps to speed variations of the tape recorder reproducing the timecode) will normally result in similar changes in the timing of MTC messages. The 'tightness' of lock between the sequencer reading the MTC data and the original timecode will depend on the software. It may be possible to select a form of 'soft' lock mode, in which small variations in the timecode rate due to jitter or wow and flutter are smoothed out. The problems are greater when MTC is used to lock digital audio replay, as discussed in section 7.6.7.

The computer's MIDI interface may also have a SMPTE/EBU timecode *output* fed from an internal timecode generator, which can be used to 'stripe' tapes with timecode under control of the computer. The timecode reader in the MIDI interface may have capabilities for 'flywheeling' over short breaks in the timecode (in other words continuing to produce MTC data at an estimated rate in the absence of a SMPTE/EBU input). This can be useful for working with tapes where there are dropouts in the timecode track. The timecode reader may also be able to 'jam sync' the generator so that new code can be generated based on the values of the old code.

In systems where the MIDI interface connected to the computer does not have timecode ports there are a number of alternative options. Figure 7.10 shows the use of an external SMPTE/MIDI convertor which takes in timecode from the external

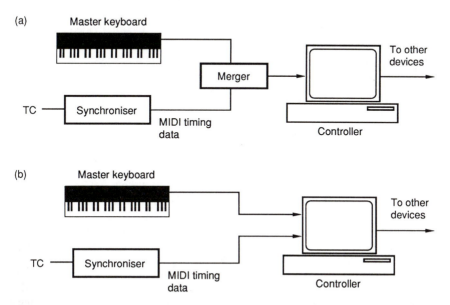

Figure 7.10 Two examples of synchronised operation. (a) Using a MIDI merger to combine MTC and master keyboard data when only one MIDI input is available on the computer, and (b) using a second MIDI input for MTC

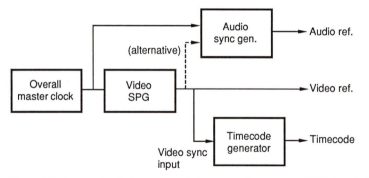

Figure 7.11 In larger installations, a central video sync pulse generator (SPG) is locked to a stable clock source. This clock source may also be used to lock a digital audio reference signal generator. The video sync signal should be used to lock the timecode generator used to stripe tapes in the studio

recorder and outputs MTC data over a MIDI interface. This synchronisation data would normally have to be connected to the computer's MIDI input. In the case of systems with only a single MIDI input to the computer it will be necessary to merge the MTC information with MIDI data coming from other places in the system using an external 'MIDI merger', in order that the timecode source does not monopolise the computer's MIDI input. Alternatively, some SMPTE/MIDI convertors have a MIDI input and an internal merging function. This would be used, for example, when using a sequencer to record information from a master keyboard whilst the replay of previously recorded tracks was synchronised to external timecode.

It is also possible to convert VITC (vertical interval timecode) into MTC data using a dedicated VITC convertor (none of the currently available computer MIDI interfaces are provided with VITC readers). Mark of the Unicorn's Video TimePiece is one example of such a convertor. As discussed in Chapter 6, VITC is stored in unseen lines of a video signal and is used for providing a timecode signal from a video machine even when it is in paused or still frame modes. This would be useful if you were intending to do a lot of work with video, where accurate frame-by-frame cueing of the picture was required, although it is important to check that the sequencer software is capable of following external MTC values that are 'nudged' in video slow motion modes.

Where digital audio or video are concerned it may be necessary to consider external synchronisation of the timecode generator if it is to be used to stripe external recordings. This is because the rate of timecode should be perfectly locked to the frame rate of the video picture or the sampling rate of the digital audio – in order that the two do not drift with relation to each other, resulting in sync slips. Such lock is normally achieved by connecting a video or audio sync signal to the timecode generator's reference input, as shown in Figure 7.11. In a large studio or broadcast environment the composite video sync signal would normally be provided by a central 'house sync' reference generator, to which all the devices in the studio would be locked, but in smaller situations it may be necessary to derive this signal from the video recorder itself.

It is less common for timecode equipment to provide a digital audio sync input, but there are a number of ways in which this problem may be solved. If such a sync input does exist it will typically be in the form of a word clock signal on a BNC connector (a square wave signal at the sampling frequency), or (very rarely) an AES-11 format sync signal (a signal in the AES/EBU audio interface format on an XLR connector). The audio sync signal should be the same signal as is used to lock the digital audio recorder, or it may be derived from the recorder's own word clock output. Without a digital audio sync input one must use a video sync source to lock the timecode generator and the audio recorder. Many professional digital audio recorders have a composite video sync input which can be used to lock the sampling frequency.

Professional video and audio recorders often have their own built-in timecode generators, and this is often a useful way of striping tapes with timecode because the internal logic of the tape recorder can be set so that the generator is locked to its own audio or video frame rate. This eliminates the need to worry about referencing an external timecode generator. It not normally necessary to be concerned with external sync of timecode generators when working only with analogue tape recorders in a music studio, because such machines are normally free running. An analogue tape can be striped with free-running timecode prior to a session, and during the session the MIDI equipment may be locked to the free-running replay of that timecode track from the tape. In general, it is advisable to stripe tapes with timecode *before* recording anything else, in order that there is then a replayed timecode signal for MIDI equipment to lock to during the rest of the session. Timecode can be post-striped, though, provided that the recorder allows timecode to be recorded separately, but it must be locked to the frame rate of the material which has been recorded.

It is vital that all the devices in a system which use timecode operate at the same frame rate. This will normally be the television frame rate of the country in which you are working (see Chapter 6), although occasionally it may be necessary to operate at other rates if work is being carried out for foreign markets.

7.6.6 Controlling external studio machines

MIDI Machine Control (MMC) is a growing possibility for integrating studio equipment into the MIDI system, as discussed in Chapter 6. In conjunction with MTC, MMC can be used to interface audio and video tape and disk recorders, as well as synchronisers and other equipment, provided that these machines are fitted with MIDI interfaces and have the appropriate internal firmware for MTC and MMC. At the time of writing these options are only available in a limited number of machines.

The system in Figure 7.12(a) shows a multitrack audio recorder connected to the computer as one of the devices linked to its multiport MIDI interface. MMC commands are issued by the sequencer running on the computer so as to link the transport functions on the tape machine to those of the sequencer, and timecode is read from the tape machine to determine its position, using the timecode port on the MIDI interface. A slightly more advanced system in Figure 7.12(b) shows that a tape recorder provided with an internal SMPTE/EBU timecode reader and a MIDI output could send MTC and MMC response data directly back to the computer without using the SMPTE/EBU timecode reader in the interface at all.

7.6.7 Systems incorporating digital audio

The most straightforward way of adding digital audio multitrack recording capabilities to a MIDI system is to install a suitable sound card in the computer or to make

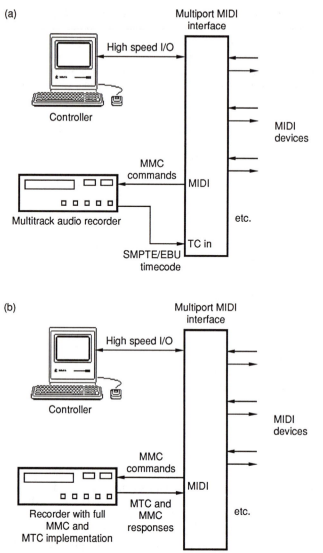

Figure 7.12 (a) Here a tape recorder is controlled using MMC commands from the sequencer, with positional information being fed back to the sequencer via the SMPTE/EBU timecode reader in the MIDI interface. (b) A tape recorder with a built-in timecode reader and a full MMC/MTC implementation does not require a timecode reader in the MIDI interface, since positional information is returned to the sequencer in the form of MTC messages

use of internal audio DSP facilities within the computer (see section 5.9). A number of sequencer packages now integrate the handling of digital recording and editing operations. The alternative is to use an external digital multitrack machine synchronised using timecode and possibly controlled using MMC, as described in the previous two sections.

Figure 5.31 showed an example of a system in which a four-channel digital audio card had been installed in an expansion slot on a computer. In such a system an external interface handles analogue and digital audio inputs and outputs, connected via a high speed link to the expansion card. A large disk drive, attached either directly to the audio card or to the computer's SCSI interface, stores the PCM audio data. Audio signals from an external mixer could be patched to any of the four inputs of the digital audio interface, and these could be derived from microphones or other sources of 'live' audio information. Depending on the software running on the computer, audio could be recorded at the same time as MIDI tracks were being recorded.

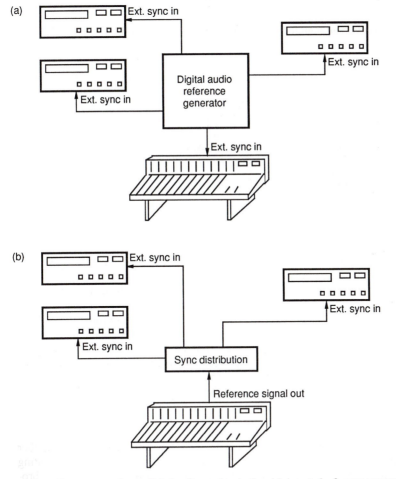

Figure 7.13 Two approaches to digital audio synchronisation. (a) A central reference generator is used to lock the sampling rates of all devices in the studio to a common clock. (b) A digital mixer operating in internal sync mode provides a reference signal for the other devices

Even though the means of recording digital audio on the computer hard disk is digital, analogue inputs and outputs are often provided in addition to digital interfaces so that the system can be connected to analogue mixing and recording equipment. It is much more straightforward to interconnect the audio signals in the analogue domain, and this will be the only option if all the external audio equipment is analogue, but the possibility exists for interconnections to be made entirely in the digital domain using standard digital interfaces such as the AES/EBU. Using digital interfaces the quality of the audio signal can be maintained, but the synchronisation issues become much more complicated. Most of this topic is outside the intended scope of this book, and has been covered in detail in *The Digital Interface Handbook* by Rumsey and Watkinson (see Appendix 2). The biggest problem in much of today's low cost digital audio equipment is either that it cannot easily be synchronised to an external clock signal, or that each device requires a different type of external clock. The key to success is to ensure that all devices which are to be interconnected digitally are locked to a common sample frequency clock provided from one central device (which could be a digital mixer), as shown in Figure 7.13. If video is involved then the audio sample rate should also be locked to the video frame rate, as described above.

The most basic form of replay synchronisation used in integrated MIDI and digital audio packages is simple triggering of the digital audio replay at appropriate points in the song. When a particular timecode value is received from an external sync source, or at the programmed MIDI beat number, the digital sound file replay is triggered. Thereafter its replay free-runs, locked to the internal crystal-controlled sampling rate. The problem is that drift can build up between the MIDI tracks and the digital audio because they are not permanently locked to the same sync source.

Alternatively, digital audio replay may be locked to an external MIDI sync source in the same way as MIDI replay. In other words, either MIDI beat clock or MTC may be used to reference the speed of digital audio replay. This is useful when synchronising an integrated digital audio and MIDI recording package to an external video recorder. Although MIDI sync of digital audio replay is possible, it will not result in the best sound quality if the digital audio replay is 'hard locked' to MTC, and it raises the additional problem that any speed variations in the MIDI sync signal (which would normally result in slight tempo changes on the MIDI sequence) could result in sample rate and therefore pitch changes of the audio. If the changes in sample rate are large it may be difficult to interconnect the digital audio output of the system to other digital equipment because the sample rate will swing outside the capture range of the receiving device. A sample rate convertor or digital audio synchroniser would be required in such a case, so as to maintain the digital audio output at a constant rate no matter what the changes in the replay rate.

An interesting option designed to provide a workaround to the problem of MIDI and digital audio sync replay is offered by Opcode in its StudioVision package. Assuming that the replay of StudioVision's audio and MIDI data is to be locked to a video tape using MTC, the operator must first play completely through the videotape with Vision in a calibration mode. The software measures the rate of the replayed timecode with reference to the audio sampling rate, noting both the maximum and minimum deviations and the long-term average rate. It then adjusts the relative timing of MIDI and audio events in the sequence to compensate, and ignores the short-term variations in timecode rate coming from the VTR.

If audio, video and timecode rates are correctly locked together throughout the operations of striping, recording and replay, it should be possible for the digital audio replay to be released from MIDI synchronisation control after initial lock is established on replay. Unfortunately many low cost MIDI-integrated systems are not versatile enough to make this possible, since it requires that the digital audio replay can be automatically switched over to external word clock or video sync after the initial lock-up period. The best audio quality is nearly always obtained when it is replayed locked to its internal crystal speed reference. In most cases when external sync is not required the sequencer replay will be locked to the computer's internal timer and the digital audio to its crystal clock, and the drift will be minimal.

7.6.8 Integrated digital video

It was indicated in Chapter 5 that random access digital video could be added to the central computer of a MIDI system, in addition to digital audio, using data reduction techniques to limit the amount of data storage and transfer bandwidth required. The basic principles have already been outlined, but it will simply be mentioned here that onboard digital video storage can be integrated successfully with MIDI and digital audio recording and replay, in order to eliminate the need for an external video recorder (except to dump the video material on a one-time basis). This would simplify the synchronisation issues considerably since the computer software would take care of the lock between all three types of program material, and there would be no waiting for lock-up and no timing variations to worry about.

7.7 Device IDs in a MIDI system

Many of the universal system exclusive message protocols include a device ID used to indicate which device is addressed by the message. This differs from the MIDI receive channel number(s) of a device, and system exclusive messages do not have channel numbers in any case. The device ID can be anywhere between 0 and 127, since it occupies a whole MIDI data byte. It is normally set on the front panel of the device concerned, using one of the system menus in which global setup options are fixed. Alternatively it may be possible to alter it using a manufacturer's SysEx message.

In a system where all devices are connected to a computer on the same MIDI bus, either using a THRU daisy chain or a star configuration from a single MIDI port (see above), the setting of device IDs will be important as a means of distinguishing between devices. A controlling computer wishing to send a universal SysEx message to one of the devices would address it using its device ID. If more than one device was set to the same ID then they would both respond, and this would not normally be desirable. When multiport interfaces are used the device ID is perhaps less important, especially when each device is connected to its own unique MIDI port. In such cases the controlling software could distinguish between devices by physical port number. It is nonetheless good practice to get used to setting unique device IDs in a large MIDI system, since many librarians and other software packages can make use of the data.

7.8 High speed networks

Multiport MIDI interfaces solve a lot of the problems associated with large systems, and expand the number of channels available above the basic sixteen. A high speed network is the next step in system integration where a number of MIDI devices may be separated by larger distances than can be managed with ordinary MIDI cables, or when it is necessary to transfer data for more channels than can be accommodated using standard computer serial interfaces such as RS422 and RS232. Such an approach could be used, for example, in large live performance rigs or for multimedia events in environments such as theme parks.

A number of computer network approaches exist in the general computing world. Their purpose is mainly to link a number of machines in order that files can

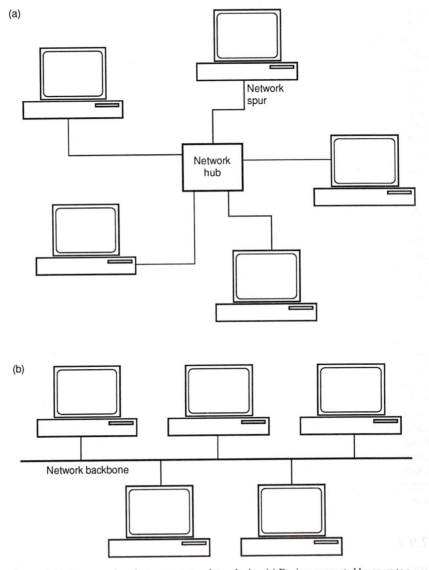

Figure 7.14 Two examples of computer network topologies. (a) Devices connected by spurs to a common hub, and (b) devices connected to a common 'backbone'

be transferred or shared between them, and in order that each machine can access common facilities such as printers, filestores, communications and input devices. The widely used Ethernet standard which runs over copper coaxial cables at 10 Mbit/s is one well known system, and the recent FDDI standard (running at 100 Mbit/s over optical fibre) is another. In a local area network (LAN), an example of which is pictured in Figure 7.14, each device has a unique address and data is transferred from one device to another in the form of packets, each packet being headed by the address of its destination. The process is not unlike that involved in MIDI messaging, except that data can be transferred bidirectionally and at rather higher speed than with MIDI.

Without going into the finer details of computer network operation it is sufficient to say here that for the purposes of carrying control information for real-time applications such as MIDI (and also audio and video data) it is important to adopt a network protocol whose performance is deterministic and real time. In other words, the maximum delay in transferring a packet of data is known and within certain limits appropriate to the application. Ethernet is not particularly suitable here because it does not fulfil these requirements very well. When the network gets busy the whole thing slows down. For this reason, the work which has been going on with relation to multimedia networks has taken a different route, defining high speed network protocols which are specifically intended for carrying real-time audio, video and control data over optical fibres.

Lone Wolf's MediaLink network is a leading example of such technology. MediaLink is a real-time optical fibre network onto which can be placed packets of data for all sorts of different multimedia applications, and the company makes devices called 'taps' which are basically gateways onto the network for different types of data. One particular tap is available for MIDI interfacing and it is possible, using a number of MIDI taps, to transfer many thousands of MIDI channels over the network without incurring noticeable delay (the speed of transfer is many megabits per second, as opposed to MIDI's 31.25 kbit/s). The company provides configuration software to set up different network configurations for different applications. Because the network uses optical fibres, distances of over a kilometre can be covered without difficulty.

Readers interested in the progress towards standardisation in real-time networks for the control of sound systems are referred to the work of the Audio Engineering Society (AES) SC-10 standards working group. This group is defining a standard means of networked communication between devices in a sound system, which may include MIDI equipment but which is not limited to MIDI equipment. A draft specification is being published, called Draft AES-24ID (see Appendix 2). The group's aim is to produce a real-time control protocol and transport system which is sufficiently reliable to be used for live applications, including the control of amplifiers, loudspeaker systems, MIDI equipment and other audio devices. It is likely to adopt already existing network technology as a means of transporting the control messages around the system, and at the time of writing the SC-10 group is considering submissions from a variety of sources.

7.9 Troubleshooting a MIDI system

When a MIDI system fails to perform as expected, or when devices appear not to be responding to data which is being transmitted from a controller, it is important to adopt logical fault-finding techniques rather than pressing every button in sight and starting to replug cables. The fault will normally be a simple one, and there is only a limited number of possible causes. It is often worth starting at the end of the system nearest to the device which exhibits the problem and working backwards towards the controller, asking a number of questions as you go. You are basically trying to find out either where the control signal is getting lost or why the device is responding in a strange way. The old saying: 'If it aint broke, don't try to fix it' is a good one. Many computer-controlled MIDI systems stay working properly because of a cocktail of good luck and perseverance with sorting out a combination of software and hardware that works together.

7.9.1 Device not responding?

Look at the hints in Figure 7.15. Firstly, is MIDI data getting to the device in question? Most devices have some means of indicating that they are receiving MIDI data, either by a flashing light on the front panel or some other form of display. Alternatively it is possible to buy small analysers which in their simplest form may do something like flashing a light if MIDI data is received. If data is getting to the device then the problem is probably either within the device or after its audio output. The most common mistake that people make is to think that they have a MIDI problem when in fact they have an audio problem. Check that the audio output is actually connected

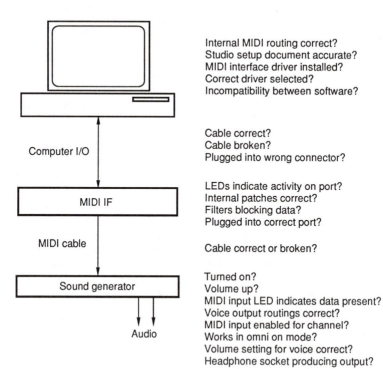

Figure 7.15 In this diagram are shown a number of suggestions to be considered when troubleshooting a MIDI system

to something and that its destination is turned on and faded up. Plug in a pair of headphones to check if the device is responding to MIDI data. If sound comes out of the headphones then the problem most probably lies in the audio system.

If the device is receiving MIDI data but not producing an audio output, try setting the receive mode to 'omni on' so that it responds on all channels. If this works then the problem must be related to the way in which a particular channel's data is being handled. Check that the device is enabled to receive on the MIDI channel in question. Check that the volume is set to something other than zero, and that any external MIDI controllers assigned to volume are not forcing the volume to zero (such as any virtual faders in the sequencer package). Check that the voice assigned to the channel in question is actually assigned to an audio output which is connected to the outside world. Check that the main audio output control on the unit itself is turned up. Also try sending note messages for a number of different notes – it may be that the voice in question is not set up to respond over the whole note range.

If no MIDI data is reaching the device then move one step further back down the MIDI signal chain. Check the MIDI cable. Swap it for another one. If the device is connected to a MIDI patcher or router of some kind, check that the patcher input receiving the required MIDI data is routed to the output concerned. Try connecting a MIDI keyboard directly to the patcher input concerned to see if the patch is working. If this works then the problem lies further up the chain, either in the MIDI interface attached to the controller or in the controller itself. If the controller is a computer with an external MIDI interface, it may be possible to test the MIDI port concerned. The setup software for the MIDI interface may allow you to enter a 'Test' mode in which you can send unspecified note data directly to the physical port concerned. This should test whether or not the MIDI interface is working. Most interfaces have lights to show when a particular port is receiving or transmitting data, and this can be used for test purposes. It may be that the interface needs to be reconfigured to match a changed studio setup. Now go back to the controller and make sure that you are sending data to the right output on the required MIDI channel and that you are satisfied, from what you know about it, that the software concerned should be transmitting.

If no data is getting from the computer to the interface, check the cables to the interface. Then try resetting the interface and the computer. This sometimes re-establishes communication between the two. Reset the interface first, then the computer, so that the computer 'sees' the interface (this may involve powering down, then up). Alternatively, a soft reset may be possible using the setup software for the interface. If this does not work, check that no applications are open on the computer which might be taking over the interface ports concerned (some applications will not give up control over particular I/O ports easily). Check the configuration of any software MIDI routers within the computer to make sure that MIDI data is 'connected' from the controlling package to the I/O port in question.

Ask yourself the question: 'was it working the last time I tried it?'. If it was, it is unlikely that the problem is due to more fundamental reasons such as the wrong port drivers being installed in the system or a specific incompatibility between hardware and software, but it is worth thinking through what you have done to the system configuration since the last time it was used. It is possible that new software extensions or new applications may conflict with your previously working configuration, and removing them will solve the problem. Try using a different software package to control the device which is not responding. If this works then the problem is clearly with the original package.

7.9.2 Device responding in an unexpected way

Assuming that the device in question had been responding correctly on a previous occasion, any change in response to MIDI messages such as program and control changes is most likely due either to an altered internal setup or a message getting to the device which was not intended for it.

Most of the internal setup parameters on a MIDI-controlled device are accessible either using the front panel or using system exclusive messages. It is often quite a long-winded process to get to the parameter in question using the limited front panel displays of many devices, but it may be necessary to do this in order to check the intended response to particular MIDI data. If the problem is one with unusual responses (or no response) to program change messages then it may be that the program change map has been altered, and that a different stored voice or patch is being selected from the one intended. Perhaps the program change number in question is not assigned to a stored voice or patch at all. If the device is switching between programs when it should not then it may be that your MIDI routing is at fault. Perhaps the device is receiving program changes intended for another. Check the configuration of your MIDI patcher or multiport interface. A similar process applies to controller messages. Check the internal mapping of controller messages to parameters, and check the external MIDI routing to make sure that devices are receiving only the information intended for them.

When more than one person uses a MIDI-controlled studio, or when you have a lot of different setups yourself, virtually the only way to ensure that you can reset the studio quickly to a particular state is to store system exclusive dumps of the full configuration of each device, and to store any patcher or MIDI operating system maps. These can either be kept in separate librarian files or as part of a sequence, to be downloaded to the devices before starting the session. Once you have set up a configuration of a device which works for a particular purpose it should be stored on the computer so that it could be dumped back down again at a later date.

7.9.3 Erratic timing or lost messages

Problems with timing or lost messages are almost certainly due to overloading of the MIDI bus. The roots of this were discussed earlier in this chapter, in sections 7.4 and 7.5. It may be necessary to redesign the MIDI routing, possibly using a more advanced multiport MIDI interface with high speed communication to the computer. Alternatively, try filtering out any unnecessary control data, such as aftertouch, for as many sequencer tracks as possible, and ensure that timing data such as beat clock or MTC is not being transmitted to ordinary sound generators. A faster computer might help, although this is a somewhat drastic solution and most sequencer software is designed to work with low end machines as well as those at the high end. It may help, though, to run the most cut-down computer system possible, with all unnecessary extensions and routines disabled, in order to allow the MIDI software to run its real-time output routines without interruption.

Contact information

The following are some useful contacts for further information about MIDI:

Standards and documentation

International MIDI Association,
5316 W. 57th Street,
Los Angeles,
CA 90056
USA
Tel. +1 310 649 6434

On-line service

PAN, the Performing Arts Network, is an on-line data service which can be accessed by phone using a modem, or over the Internet if you have access to Internet services. It is based in the USA, but there are a lot of European users. PAN acts as a point of contact between people involved in MIDI, music, computer music and audio, and can be used for conferencing, accessing archived information or obtaining the latest documentation on MIDI-related subjects. It also provides hot lines to many manufacturers, developers and standards groups. It is a chargeable service, but the rates are reasonable. Details can be obtained from:

Performing Arts Network
PO Box 162
Skippack
PA 19474
USA
Tel. +1 215 584 0300
Fax. +1 215 584 1038

1200 baud modem: +1 617 576 0862
2400 baud modem: +1 617 576 2981
Internet address: pan.com

Selected further reading

The following books, articles and documentation have been selected from a wide range of available material as forming a useful further reading list:

MIDI

Conger, J. (1989) *C Programming for MIDI*. M&T Publishing, Inc. Redwood City, CA., USA

DeFuria, S. and Scacciaferro, J. (1989) *MIDI Programmer's Handbook*. M&T Publishing, Inc. Redwood City, CA., USA

DeFuria, S. and Scacciaferro, J. (1987) *The MIDI System Exclusive Book*. Third Earth Publishing, Inc., Pompton Lakes, NJ, USA

Heywood, B. and Evan, R. (1991) *The PC Music Handbook* . PC Publishing, Tonbridge, UK

MMA (1983) *MIDI 1.0 Detailed Specification*. International MIDI Association

MMA (1986) *MIDI Sample Dump Standard*. International MIDI Association

MMA (1987) *MIDI Timecode and Cueing: Detailed Specification*. International MIDI Association

MMA (1988) *Standard MIDI Files 1.0*. International MIDI Association

MMA (1991) *General MIDI System Level 1*. International MIDI Association

MMA (1991) *MIDI Show Control 1.0*. International MIDI Association

MMA (1992) *MIDI Machine Control 1.0*. International MIDI Association

MMA (1993) *4.2 Addendum to MIDI 1.0 Specification*. International MIDI Association

Huber, D. (1991) *The MIDI Manual*. SAMS, Carmel, IN., USA

Moog, R. (1986) MIDI: the Musical Instrument Digital Interface. *Journal of the Audio Engineering Society* , vol. 34, no. 5, pp. 394–404, May

Rothstein, J. (1992) *MIDI: A Comprehensive Introduction*. Oxford University Press, Oxford, UK

Russ, M. (1992) MIDI: past, present and future. *Sound on Sound* , pp. 58–66, June

Yavelow, C. (1987) Computers and music: the state of the art. *Journal of the Audio Engineering Society* , vol. 35, no. 3, pp. 161–193, March

Yavelow, C. (1992) *Macworld Music and Sound Bible*. IDG Books Worldwide, Inc., San Mateo, CA., USA

Digital audio

Rumsey, F. (1990) *Tapeless Sound Recording*. Focal Press, Oxford, UK and Boston, MA., USA

Rumsey, F. (1991) *Digital Audio Operations*. Focal Press, Oxford, UK and Boston, MA., USA

Rumsey, F. and Watkinson, J. (1993) *The Digital Interface Handbook*. Focal Press, Oxford, UK and Boston, MA., USA

Watkinson, J. (1993) *The Art of Digital Audio*, 3rd edition. Focal Press, Oxford, UK and Boston, MA., USA

Networks

AES (1993) Draft AES-24ID. AES information document for sound system control – application protocol for controlling and monitoring audio systems (AES-24). In *Journal of the Audio Engineering Society*, vol. 41, no. 12, December

Karagosian, M. (1993) Report of SC-10-1 working group on data communications. In Standards News, *Journal of the Audio Engineering Society* , vol. 41, no. 11, pp. 946–950, November

Lacas, M., Warman. D, and Moses, R. (1993) The MediaLink real-time multimedia network. Presented at the *AES 95th Convention, New York, October 7–10*, preprint 3736. Audio Engineering Society

Moses, R. (1993) Report of SC-10-2 working group meeting. In Standards News, *Journal of the Audio Engineering Society* , vol. 41, no. 10, p. 799, October

Wilkinson, S. (1993) Sound all around. *Electronic Musician*, October, pp. 70–76

Miscellaneous

Lamaa, F. (1993) Open Media Framework Interchange. In *Proceedings of the AES UK Digital Audio Interchange Conference, London, 18–19 May*, pp. 77–94. Audio Engineering Society, British Section publication

Rumsey, F. (1993) Digital video bit-rate reduction. *Studio Sound*, October, pp. 83–90

Index

THE DIGITAL
INTERFACE HANDBOOK

FRANCIS RUMSEY AND JOHN WATKINSON

- Points out procedures to take if interface fails to work
- Details what sort of test equipment to employ
- Covers audio and video interfacing principles

The Digital Interface Handbook is a thoroughly detailed manual for those who need to get to grips with digital audio and video systems. Now that installations in the broadcasting, multimedia and music industries are increasingly 'all digital', engineers and operators working in these industries need to become more familiar with digital interfaces, their benefits and pitfalls. Digital interfaces are the key to maintaining programme quality throughout the signal chain.

In **The Digital Interface Handbook** Francis Rumsey and John Watkinson bring together their combined experoence to shed light on the differences between audio interfaces such as AES/EBU, SPDIF, MADI and other manufacturer-specific implementations, showing how to make devices 'talk to each other' in the digital domain despite their subtle differences. They also include detailed coverage of all the regularly used digital video interfaces. Anyone who has spent half a day wondering why two tape recorders will not communicate needs this book!

0 240 51333 9   224PP   PAPERBACK   APRIL 1993

SOUND AND RECORDING
AN INTRODUCTION

FRANCIS RUMSEY
TIM McCORMICK

Designed as an easy-to-read reference for those at an early stage in their careers. Especially relevant for students or trainee engineers entering music recording, broadcasting or associated industries. This book will provide a vital introduction to the principles of sound, perception, audio technology and systems.

Contents: What is sound?; Auditory perception; A guide to the audio signal chain; Microphones; Loud-speakers; Mixers; Basic operational techniques; Analogue tape recording; Noise reduction; Digital recording; Record players; Power amplifiers; Lines and interconnections; Outboard equipment; Timecode and synchronisation.

320PP   0 240 51313 4   234x156MM   PAPERBACK   1992

THE ART OF DIGITAL AUDIO
Second edition

JOHN WATKINSON

The first edition of this book is regarded as a classic in its field. Now completely rewritten to reflect the enormous recent advances in the subject it is even more comprehensive, now covering practical devices such as DCC and MiniDisc and including totally new treatments of principles such as:

• oversampling • data reduction • noise shaping • DAB • dither

This new edition begins with a chapter which is almost an introductory book in its own right and makes the remainder of the work accessible to all. At the other end of the spectrum more references than ever are included to permit serious study. What has not changed is the approach. Every subject is explained from first principles, because if the mechanism is understood it can be applied to many problems. This is an introductory, theory, applications and reference book all in one.

John Watkinson is an independent consultant in digital audio, video and data technology. He is a fellow of the AES, is listed in Who's Who in the World and is the author of five other Focal Press titles, including The Art of Digital Video - the definitive work in its field. He is also co-author, with Francis Rumsey, of the Digital Interface Handbook.

CONTENTS:

WHY DIGITAL? SOME ESSENTIAL PRINCIPLES; CONVERSION; ADVANCED DIGITAL AUDIO PROCESSING; DIGITAL RECORDING AND TRANSMISSION PRINCIPLES; ERROR CORRECTION; TRANSMISSION AND BROADCAST SYSTEMS; ROTARY-HEAD RECORDERS; STATIONARY-HEAD RECORDERS; MAGNETIC DISK DRIVES; DIGITAL AUDIO EDITING; OPTICAL DISKS IN AUDIO; DIGITAL AUDIO IN VIDEO TAPE RECORDERS.

REVIEWS OF THE FIRST EDITION INCLUDE

'Once in a while, someone writes a technical book which becomes the definitive work or industry bible. John Watkinson has done just that with The Art of Digital Audio'.
New Scientist

'Mr Watkinson's work commands pride of place amongst the publications on digital audio techniques.'
EBU Review

'... a must for all radio engineers, interested in complete digital fluency.'
Radio World

0 240 51320 7   704PP   HARDBACK   DECEMBER 1993

AVAILABLE FROM ALL GOOD BOOKSELLERS OR IN CASE OF DIFFICULTY PLEASE PHONE OUR UK DIRECT ORDER LINE ON (0933) 410511 WITH YOUR CREDIT CARD DETAILS READY OR IN THE USA PHONE 800-366-2665.

THE SOUND STUDIO
Fifth edition

ALEC NISBETT

Formerly **The Technique of the Sound Studio**, this fully revised, updated and redesigned fifth edition brings this classic text right up-to-date. It is for everyone who maintains a creative interest in sound, whether as a sound recordist, balancer, audio engineer or as a director, writer, performer or student.

Each theme is carefully discussed in non-technical terms, enabling the reader to get to grips with any and every subject in this area.

Features of this new edition include:

- all the latest developments in the audio field
- extensive coverage of the increasing use of digital systems
- more on commercial studio techniques

CONTENTS:

AUDIO TECHNIQUES AND EQUIPMENT; THE SOUND MEDIUM; STEREO; STUDIOS; MICROPHONES; MICROPHONE BALANCE; SPEECH BALANCE; MUSIC BALANCE; MONITORING AND CONTROL; VOLUME AND DYNAMICS; FILTERS AND EQUALIZATION; ARTIFICAL REVERBERATION; LINE SOURCES; FADES AND MIXES; SOUND EFFECTS; SHAPING SOUND; TAPE EDITING; FILM AND VIDEO SOUND; PLANNING AND ROUTINE; COMMUNICATION IN SOUND; GLOSSARY; BIBLIOGRAPHY.

REVIEWS OF THE PREVIOUS EDITION INCLUDE:

'Must teach even the professional quite a few new ideas ... an authoritative book.'
Studio Sound

'... the book is of particular value to the advanced amateur and also to the beginning professional.'
The Journal of the SMPTE

0 240 51292 8   480PP   PAPERBACK   APRIL 1993

AVAILABLE FROM ALL GOOD BOOKSELLERS OR IN CASE OF DIFFICULTY PLEASE PHONE OUR UK DIRECT ORDER LINE ON (0933) 410511 WITH YOUR CREDIT CARD DETAILS READY OR IN THE USA PHONE 800-366-2665.

THE USE OF MICROPHONES
Fourth edition

ALEC NISBETT

The Use of Microphones includes, in a succinct way, all you need to know about how to choose the right microphones for the job, how to position them, move them, balance and control them to make sure of the sound quality you need. Now completely updated to include more on radio microphones, contact and miniature microphones and combination stereo microphones this fourth edition continues the tradition of being an easy to read guide for students or professionals who wish to expand their skills.

PARTIAL CONTENTS:

LISTENING TO SOUND; WAVELENGTH; FREQUENCY; WAVES AND PHASE; MUSICAL ACOUSTICS ; THE HUMAN VOICE; LISTENING CONDITIONS; MICROPHONE CHARACTERISTICS; WIND, WATER AND MICE; BALANCE; SOUND AND PICTURE; MUSIC BALANCE; CONTROL; USEFUL FORMULAE.

0 240 51365 7   192PP   PAPERBACK   JANUARY 1994

AVAILABLE FROM ALL GOOD BOOKSELLERS OR IN CASE OF DIFFICULTY PLEASE PHONE OUR UK DIRECT ORDER LINE ON (0933) 410511 WITH YOUR CREDIT CARD DETAILS READY OR IN THE USA PHONE 800-366-2665.